I0050244

Understanding the universe of telecoms and ICTs

Understanding the universe of telecoms and ICTs

Gregory Domond

By the same Author
The power of your words
Servir pour être Grand
Négligence, votre premier ennemi : comment s'en débarrasser
Va avec cette force que tu as
Manuel d'évangélisation
Dieu est un Dieu de détails

Legal Depot
ISBN: 978-99970-53-00-8
© No part of this publication may be reproduced in any form whatsoever without the prior permission of the author

ACKNOWLEDGEMENTS

I thank all those who have read and commented on the articles on telecommunications and ICTs published in the columns of the newspaper Le Nouvelliste. Their encouragement has prompted me to publish these articles in this form to serve more readers around the planet. To them I am indebted for this new publication.

My thanks go to my friend, Edva ALTEMAR, for his encouragement throughout the drafting process, and the preface of the French version that he wanted to write.

Mr. Joses JEAN BAPTISTE deserves special thanks for the preface to the English version he provided.

Thank you in advance to the new readers for their appreciations and comments. They are the reason of this manual.

All the glory is to my God who distributes all capacities.

Table of Contents

PREFACE TO THE ENGLISH VERSION

After the first three volumes, Engineer Gregory DOMOND is offering the public his fourth technology book entitled " *Understanding the universe of telecoms and ICTs*".

The first three books written by Mr DOMOND require basic knowledge in telecommunications as pre-requisite. The reader must be a professional in Telecoms/ICTs or in a related field. It is not the same for the book: *Understanding the universe of telecoms and ICTs*. The book is not centered on a specific topic. It's rather an introduction to telecommunications and ICTs industry. The author presents basic concepts for different areas such as broadcasting, radiocommunication, telephony, data transmission, Internet, mobile telephony, broadband services, etc..

The definitions are clear and concise, the historicity of the services is precise. This is, in my opinion, the merit and the originality of this book. With this manual, knowledge in the telecoms field is within the reach of everyone. Moreover, for each basic service, the author also presents the associated services or associated concepts. So, the basic knowledge is widened and expanded.

Certainly, the consumer and beginner will understand and appreciate the methodology. But not only the beginners. Student in Telecoms and ICTs and related fields like informatics, networking will too! The book is a quick reference for telecoms/ICTs terminology.

Understanding the universe of Telecommunications and ICTs, a book for beginner or a telecom dictionary? The answer is both. It's a book to read and to have in your library!

Joses JEAN-BAPTISTE, Eng. M.Sc

Professor of Telecommunications

PREFACE TO THE FRENCH VERSION

The dizzying development of telecommunications and Information and Communication Technologies (ICTs) elicits an incredible curiosity and thrust from users who seek to understand this growing industry, which puts at their disposal so many means of remote communication. Telecommunications, which once were reserved to insiders, are today accessible to all. Their conquests have, in a few decades, revolutionized the companies all over the world, to the extent of making a global village. A total change is observed in all spheres of social activities, and it calls everyone to understand the progress, the evolution and the new projections of telecommunications. This essential approach requires a basic knowledge of the key concepts, an understanding framed with the current reality and which considers the past, while pointing toward the future.

From telegraph to mobile telephony, mutations in the telecommunications and in Information and Communication Technologies (ICTs) sector are multiple. It is followed by a technological progression almost exponentially. Telephony from 1776 to 1980 that was considered fixed, has known from 1980 to date five generations of mobile telephony. Computing progresses, converges and merges with telecommunications in order to suggest the use of a quantity of data so large that it is qualified of mega data. The Internet is no longer confined to the computers and human connections, but the network of networks is extended to the connection of objects, thus making the urban activities more and more independent, and they give birth to the smart cities. The appropriation of telecommunications, even from the point of view of a user, requires an acute knowledge of the related jargon, terms and concepts. The professor of telecommunications, Gregory DOMOND, aware of the need, has developed "*Understanding the Universe of Telecommunications and ICTs*" to bring you this knowledge of the concepts in order to help a better understanding of the world of telecommunications, why not the information and communication technologies (ICTs). Entering into all the details of the terms, the document bears its name well, since telecommunications and ICTs are a universe to be explored by all consumers of services, and all the other stakeholders in this so dynamic a sector. The reader will find in addition to the definition of key concepts, the different resources used in telecommunications, the environment of the sector, the services available, the techniques and technologies in use. Engineer DOMOND provides explanation about telephony, broadcasting, computing and the Internet in a language that is adapted to all. A little further, the author and technological columnist details the economic issues related to the telecommunications sector, which is the second largest economy in the world.

Developed under a format of presentation, the reader of the book will not get tired seeking to link the chapters. There he will draw according to its shortcomings, his area of intervention, and his preferences of subject. The consumer, the professional, the student, the researcher and the interested parties will find a tool adapted to their needs and will be able to unequivocally sharpen their knowledge of the telecommunications sector. This book is intended to all stakeholders (users, professionals, providers of services, investors) of the telecommunications and ICTs chain, wishing to better understand this industry.

I recommend this document to all those who want to get a better comprehension of the sector, by doing this; they will abandon amateurism for the benefit of professionalism.

Edva ALTEMAR
Telecommunications and social sciences professional
Researcher at the Research and Study Group in Telecommunications and Information Systems (GRETSI)

SUMMARY

The book entitled "*Understanding the Universe of Telecommunications/ICTs*" intends to help all stakeholders, including consumers, to better understand the telecommunications sector for the purposes of an optimal use, and a perspective on future developments. The document developed in the form of a presentation is composed of 9 chapters.

Chapter 1, "*Introduction to telecommunications and resources used* " defines the concepts of telecoms and ICTs, explains the different types of electronic communications. This chapter briefly presents the 7 resources vital to the provision of telecommunications services: Radio frequency spectrum, telephone numbering plan, Internet domains, IP addresses, high points, existing infrastructures and satellite orbits. Their management and uses are briefly explained in this first part.

Chapter 2 "Telecom/ICTs Environment" deals with the framework for the operation of the sector. It presents, among other things, the different stakeholders of the chain, the ecosystem, the legal and regulatory aspects, and the standardization of telecommunications.

The telecommunications Services and their uses are studied in the third chapter. The principles and modes of operation of the major services are presented in a simplified manner.

The fourth part presents techniques and technologies used to design, develop and provide telecoms services. The concept of "network" is succinctly developed.

Chapter 5 is dedicated to telephony. It reviews the different techniques and technologies related to fixed and cellular telephony. The different services associated with telephony as well as the means of access are presented there.

Chapter 6 is devoted to "Sound and television broadcasting". The techniques and means of access to radio and television are presented. The need for and the process of transition from analogue television to digital television are explained in this chapter.

Chapter 7 "Computing and the Internet" provides the basic notions relating to computing and the Internet. It studies mostly the techniques, technologies, services and means of access to the Internet.

The economic aspects of the telecommunications sector, which is the second world economy, are discussed in chapter 8.

Chapter 9 deals with the use of telecommunications to manage natural disasters.

The reader of the book will have an overview of the telecommunications/ICTs sector, and will have practical information for a better use of derived services. This document can generate among the young people interests and vocations for telecommunications and ICTs, the foundation of the information society in accelerated construction.

CHAPTER 1

INTRODUCTION TO TELECOMMUNICATIONS AND USED RESOURCES

DEFINITION OF TELECOMMUNICATIONS

- ✓ Any transmission, emission or reception of signs, signals, writings, images and sounds or intelligence of any nature by wire, radio, optical or other electromagnetic systems[1]
- ✓ Set of the technical means necessary for the routing of information as faithful and reliable as possible between two points a priori whatsoever, at any distance, with reasonable costs[2]
- ✓ Set of technologies, devices, equipment, facilities, networks and applications facilitating remote communication
- ✓ Communication by wire, radio, optical or other electromagnetic means

Telecommunications or electronic communications

Telecommunications and electronic communications: 2 interchangeable concepts

Telecommunications: Remote Communications facilitated by electronic means and devices

Electronic communications

- ✓ Process allowing exchange of information between human beings, between humans and machines, and between machines
- ✓ Interactions between signals and electronic systems
- ✓ Action of a message (signal) on an electronic system
- ✓ Telecommunications: Man-to-man Dialog (exchange of information), through a machine or a system
- ✓ Exchange of information (in the form of a signal) between a transmitter and a receiver using a transmission channel
- ✓ Communication through networks
- ✓ Remote communication facilitated by electronic means
- ✓ Exchanges of information made electronically (by opposition to the transmission of objects by the traditional mailing in physical form)
- ✓ Transfer of information between two devices

1 Terms and Definitions https://life.itu.int/radioclub/rr/art1.pdf

2 Systèmes de télécommunications - base de transmission par P.-G. Fontolliet

Practical definition of Informations and Communication Technologies (ICTs)

- ✓ Technologies and equipment responsible for access, creation, collection, storage, transmission, reception, dissemination or distribution of information and communication[3]
- ✓ Technologies used to transmit, store, create and share or exchange multimedia information

Origins of telecommunications (electronic communications)

- ✓ Telecommunication = Tele + Communication
- ✓ Tele: Greek prefix meaning ***far***
- ✓ Communicare: Latin word meaning ***to share***
- ✓ Telecommunications: remote sharing of information
- ✓ Telecommunication: Remote communication facilitated by electronic means
- ✓ Term used for the first time in 1904 by Edouard Estaunié, Engineer at Posts and Telegraphs, Director from 1901 to 1910 of the School of Posts and Telegraphs, to designate the different networks put in place for the dissemination of written and sound signals

History of Telecommunications

- ✓ *18th century and Event*

  1792: Beginning of the optical telegraph of Claude Chappe

- ✓ *19th century and Events*
  - o 1832: Invention of the electric telegraph by Samuel Morse
  - o 1854: Telephone project of F. Bourseul
  - o 1860: Laws of electromagnetism by Maxwell
  - o 1865: Creation of the International telegraph Union (ITU)
  - o 1866: First Trans-Atlantic telegraph cable

3 REPORT on the work carried out by the Correspondence Group on – the Elaboration of a Working Definition of the Term "ICTSSS" – https://www.itu.int/md/dologin_md.asp?lang=fr&id=S14-PP...

- 1874: Electric Telegraph of Emile Baudot (teleprinter)
- 1876: The telephone of Graham Bell & Elisha Gray
- 1876: First recordings of Thomas Edison
- 1887: Radio waves of H. R. Hertz
- 1892: Automatic Telephone of Almon Strowger
- 1892: Broadcasting by William Crookes
- 1896: First TSF link by Guglielmo Marconi
- 1897: Radio transmission at the Pantheon of Paris by Eugène Ducretet
- 1898: Establishment by Camille Tissot of the first French operational radio link at sea.
- 1900: Induction coils of Pupin
- 1900 : Equipment of the national navy of its first devices of TSF by Camille Tissot

✓ *20 th century and Events*

- 1901: First Trans-Atlantic radio connection
- 1904 : Radio links established by The Ouessant TSF station of Camille Tissot on 600 meters with a fleet of 80 cruise liners
- 1906: Diode of Sir John Ambrose Fleming
- 1906: Triode-like "Audion" of Lee of forest
- 1912: Transmission of text by Édouard Belin
- 1914: Moving Images of Georges Rignoux
- 1915: Automatic Telephone Rotary
- 1917: Military radio of General Ferrié
- 1921: The first *carrier - current* of Edwin Colpitts and Otto Blackwell
- 1922: The first regular broadcasting programs of the Eiffel Tower
- 1925: First television company of John Logie Baird
- 1926: First electronic long-distance cable
- 1929: Kinescope of Vladimir Zvorykine
- 1932: Creation of the International telecommunications Union, ITU
- 1935: Regular television programs from the Eiffel Tower
- 1936: First Creed telex
- 1938: Principles of digitization by A. Reeves
- 1940: Creation of the CCIT, Coordination Committee of Imperial Telecommunications
- 1941: Electronic computer of G. Stilitz and Howard Aiken
- 1941: Development of the radar
- 1943: First Electronic Computer ENIAC of J. Mauchly and J.-P. Eckert
- 1947: Invention of the transistor by W. Shockley
- 1951: First microwave transmissions
- 1954: First transistor radio receiver
- 1956: Transistorized submarine cable
- 1959: First integrated circuits of J. Kilby and R. Noyce
- 1962: First satellite television link America-France from Pleumeur-Bodou
- 1966: First digital PCM link
- 1970: In France, experimentation of the first electronic time division switching system
- 1970: Optical fibers of Corning Glass
- 1970 - 1980: 1st Generation of cellular telephony (1G)
- 1971: First microprocessors
- 1972: Commissioning of the first commercial electronic switches by France and then the United States
- 1977: Operational opening of the interbank network SWIFT
- 1980: Opening of the first mobile telephone networks in Japan, then in Europe

- ✓ 1980 -1990: 2nd Generation of cellular telephony (2G)
  - o 1983: Officialization of TCP/IP as the protocol of the Internet
  - o 1987: Optical amplification by doping with Erbium
  - o 1991: Creation of the ATM Forum
  - o 1993: First SMS sent in Finland
  - o 1998: Operational use of DWDM network and foundation of 3GPP
  - o 1999: Commercialization of ADSL connections in private residences in France

- ✓ *21st century and Events*
  - o 2000: 2.5G (GPRS)
  - o 2001: Agreement of the European Union to launch the Galileo project
  - o 2003: 2.75G (EDGE)
  - o 2009: Expansion, advertising and controversy around Facebook
  - o 2009: Success of the iPhone, lower profitability of Nokia : rush to smartphones.
  - o 2000 – 2010: Beginning of 3G of cellular telephony
  - o 2011: Generalization of the Digital terrestrial television on French territory[4]
  - o 2010: 4G of cellular telephony
  - o 2015: Beginning of5G of cellular telephony
  - o 2030: Beginning of 6G
  - o 2040: Beginning of 7G
  - o 2045: Beginning of 7.5 G

Telecommunications process

- ✓ Initiation of message exchange by the sender
- ✓ Transformation of the message (conversion of the message into an electrical signal by the transmitter)
- ✓ Signal processing by the transmitter
- ✓ Transmission of the signal to the destination/to the receiver by the transmission medium)
- ✓ Signal capture by the receiver terminal
- ✓ Signal processing by the receiver
- ✓ Conversion of the electrical signal into an original message
- ✓ Reception of the message by the recipient
- ✓ Interpretation of the message by the recipient

Transmission and reception in Telecommunications

- ✓ Transmission of a message in any form
- ✓ Reception of a message faithful to the one transmitted

Levels of Transmit and Receive Powers

- ✓ Power output in Watt, kilowatt (KW), megawatt (MW)
- ✓ Power received in milliwatt (mw), microwatt (*muw*), *nanowatt*(nw), picowatt (pw)
- ✓ Cause of this low level of signal on reception: Attenuation during transmission

Challenges faced by Telecommunications

- ✓ Time: Exchange of information between two users in real time
- ✓ Distance: Exchange of information made possible between any two points of the earth's surface (whatever the distance between these two points)
- ✓ Transmission of as much information as possible on a narrow band of frequencies

Limitations of face-to-face communication

- ✓ Challenge: Coverage of a great distance
- ✓ Solution: Electronic means implemented to cover a great distance

4 Histoire des Télécommunications https://fr.wikipedia.org/wiki/Histoire_des_telecommunications

TYPES OF ELECTRONIC COMMUNICATIONS

6 types of electronic communication available

1.- Audio

- ✓ Audio electronic communication made possible thanks to the invention of the telephone in 1876
- ✓ Audio communication by AM and FM broadcasting.
- ✓ Cellular telephony, another form of electronic audio communication
- ✓ Human voice also transported over the Internet (Voice over IP)

2.- Video

- ✓ Electronic communication service synchronously combining moving images and sounds through analogue and digital television systems
- ✓ Videophone (moving image)
- ✓ Video broadcasting on the Internet

3- Text Messaging

- ✓ Text messaging: Electronic communication service provided by mobile communication networks in the form of 160 alphanumeric text messages
- ✓ Delayed or off - line service, that is, not delivered in real time
- ✓ Communication with machines and systems via SMS (eg USSD)

4- Website

- ✓ Website: Electronic communication service accessible via a web address
- ✓ Platform for communicating information in different forms: audio, texts, videos, etc.
- ✓ Virtual space for information and interaction consultations

5.- E-mail

- ✓ Exchange of multimedia messages in electronic form between two or more users having an email address.
- ✓ Electronic communication service widely used both in the formal and in the informal activities
- ✓ Highly flexible electronic communication service allowing messages to be delivered to recipients' e-mail boxes at any time

6.- Instant Messaging

- ✓ Exchange all kinds of messages with contacts almost in real time via the Internet.
- ✓ Almost instantaneous exchanges possible with online contacts[5]

Ways of electronic communications

- ✓ *Simplex or one-way communication* : Transmission of signals in one direction (transmitter to receiver)
  - o Examples: radio, television, computer addressing bus, beeper)
- ✓ *Half-duplex or Bilateral or Bidirectional Alternate communication*: alternate two-way signal transmission
  - o Example: Walkie-talkie, push to talk, Citizen Band (CB), Amateur Radio
- ✓ *Full Duplex or Bilateral communication*: simultaneous two-way signal transmission
  - o Exampl.: Telephony

Man - Machine and Machine - Man Communications

- ✓ Set of information exchange between a human user and a machine or a system

Man - Machine Communications

Communications between man and machine

- ✓ Examples: Check of the balance of the telephone account by dialing a special number (answer provided by a computer server)
- ✓ Search for information on the Internet (Information provided by computer servers)
- ✓ Consultation of the electronic mailbox (access facilitated by electronic mail servers)
- ✓ Activation of a system with an authenticated voice (Speech Recognition)

Elements used in the Man - machine communication

- ✓ Keyboard and mouse

5 Six types of electronic communication http://science.opposingviews.com/six-types-electronic-communication-1531.html

- ✓ Speech (speech recognition)
- ✓ Gesture
- ✓ Glance

Machine - Man Communication

Communication between machine and man

- ✓ Examples: Voicemail activated during a telephone call attempt
- ✓ Reception of notification of a server on the return of a message (for example: return of an email)
- ✓ Charts on screens of ticket sale machines (communication of the machine with man using signs, arrows, images, etc.)
- ✓ Automatic sending of information by a server to a list of subscribers

Elements used in machine-man communication

- ✓ Audio (voicemail)
- ✓ Video
- ✓ Graphic
- ✓ Vocal synthesis

Machine - machine communication

Communications between machines

- ✓ Direct exchanges or direct communications between devices using wired or wireless communication channels
- ✓ Examples: Communication between a sensor and an application
- ✓ Exchange of information between 2 computing servers
- ✓ Telemetry: Information collected by a device and sent to a system for analysis and decision

Elements used in machine-machine communication

- ✓ Softwares
- ✓ Sensors
- ✓ RFID (Radio Frequency Identification)
- ✓ Wi-Fi (Wireless Fidelity)

Composition of ICTs

Derived elements of the following sectors

- ✓ Computing sector (computers, servers, network equipment, etc.)
- ✓ Electronic sector (electronic components, semiconductors, receivers, televisions, etc.)
- ✓ Telecommunications sector (transmission equipment, switches, relays, terminals, etc.)

ICTs Sector Activities

- ✓ Production (Manufacturing of computers, computer hardware and telecommunications equipment, etc.)
- ✓ Distribution (Trade of ICTs and Telecoms material and equipment)
- ✓ Service provision (Telecommunications, IT, audiovisual services, etc.)

Characteristics of ICTs

- ✓ Real time: electronic exchanges supported without time difference
- ✓ Hard and soft products: combination of hardware and software
- ✓ Non-geographic: use of products and services not subject to fixed location
- ✓ Transversal: Services applicable to all sectors of activities
- ✓ Digital and virtual nature: use of digital technologies for a non-real representation of realities

Advantages of electronic communications

- ✓ Multichannel aspect: use of several channels (sound, music, image, video, text, data)
- ✓ Hypertext: link established between documents by hyperlinks

Fast transmission

- ✓ Communications possible in seconds between two very distant points of the world
- ✓ Radio broadcasts listened almost instantly
- ✓ Real-time telephone conversations
- ✓ Signal transmission at the speed of light in free space

Great coverage

- ✓ Radio coverage of an entire region by a transmission site

- ✓ Radio programs heard over hundreds of kilometers by the propagation of radio waves
- ✓ Radio coverage of one-third of the earth by a geostationary satellite

Multimedia communication

- ✓ Possibility of simultaneous exchange of multimedia information (sounds, images, videos and texts)

Low price

- ✓ Economy compared to travel expenses
- ✓ Saving of travel time

Remote work

- ✓ Possibility of remote work
- ✓ Remote operations management
- ✓ Organization of meetings via videoconference
- ✓ Work from the employee's home and electronical submission

Long-term storage and easy access

- ✓ Storage of messages exchanged from different ways: discs, magnetic tape.
- ✓ Message printing for saving the paper version
- ✓ Backup capability in cloud computing

Mobility of devices

- ✓ Contact with loved ones guaranteed through the portability of the cellular telephone and the computer
- ✓ Possibility of production/work in public areas such as: trains, cafes, airplanes, etc.

Disadvantages of electronic communications

Security of electronic exchanges

- ✓ Possible changes to messages sent by electronic means
- ✓ Possible sending of viruses and worms via email by malicious people
- ✓ Vulnerability of some user groups to cyber attacks

Addiction

- ✓ Potential dependence on ICTs through misuse of available services
- ✓ Serious consequences of cyber addiction on the health and professional activities of users of technological applications

Loss of confidentiality

- ✓ Threat caused by the interception of electronic communications
- ✓ Ability to intercept an e-mail during transit via computers and routers

Doubtful authenticity

- ✓ Possible loss of data packets during the transfer from one computer to another via a router
- ✓ In case of router overload, longer waiting time
- ✓ Dubious message with a modification noticed in the header

Opportunities of electronic communications

- ✓ Combination of sound, text, graphs, video in one message
- ✓ Interactivity of electronic communications
- ✓ Multilateral communications

Real-time electronic communications

- ✓ Exchange of information between users instantaneously or with negligible latency
- ✓ Real-time communications = live communications
- ✓ Exchange between the parties without noticeable delay
- ✓ Negligible time of processing and signal transmission by the system
- ✓ Signal propagation time and imperceptible processing time for the receiver
- ✓ Telephone conversation between two subscribers separated by a time difference
  - o Essential character of interaction and continuous feedback in telephony
- ✓ Examples: telephone conversation, videoconference, bilateral or multilateral radiocommunication, instant messaging, Internet chat, etc.
- ✓ Requirements for real-time communications:

- o Establishment of direct routes between the source and the destination;
- o Connection (status: online) of parties involved in the instant exchange

Delayed or off-line electronic communications

- ✓ Sender's messages not delivered to the destination instantaneously
- ✓ Messages delivered after a certain time varying from one case to another
- ✓ Non-urgent messages placed in the category of delayed communications
- ✓ A way to avoid the costs of real-time communication (an SMS instead of a telephone call)
- ✓ Messages stored in the system and transmitted later
- ✓ Examples: SMS, MMS, voicemail, email, etc.

Advantages of real-time communication

- ✓ Simulation of a reality close to face-to-face conversation
- ✓ Indispensable in telephone conversations, tele-education, online games, geolocation

Disadvantages of real-time communication

- ✓ Consumption of many resources of the telecommunications network (in telephony, a circuit dedicated to both speakers for the duration of the conversation)
- ✓ Impossibility for the network to serve all customers instantly[6]-[7]

Telecommunications system

- ✓ Physical and non-physical infrastructure designed to carry information from a source to a destination

System of installed and interconnected electronic equipment for the transport of electronic communication from the source to the destination

- ✓ Set of processes and equipment put in place for the transmission of information from the transmitter to the receiver
- ✓ System in charge of transporting information between a transmitter and one or more receivers connected by a communication channel, in the form of a signal

Principle of telecommunications systems

Basis of operations

- ✓ Exchanges between signals and systems (Interactions between signals and systems)
- ✓ Reactions to commands (signals) from users

Objective of a communication system

- ✓ Transmission of information from the source to the destination through a cable or free space

Components of a communication system

- ✓ Terminals (input/output devices, point of departure/arrival)
- ✓ Transmission channels
- ✓ Telecommunications processors (analog/digital conversion, control of the transmission medium)
- ✓ Control software (functionality and activity control)
- ✓ Messages to be transmitted
- ✓ Protocols

Examples of telecommunications system

- ✓ Sound system
- ✓ Email
- ✓ Instant messaging
- ✓ Video conference
- ✓ Radio and television stations
- ✓ Telephone network
- ✓ Global Positioning System
- ✓ Internet

Elements required for telecommunications systems

- ✓ Techniques
- ✓ Technologies
- ✓ Equipment (telecom, computing)

---

6 Real time communication http://searchunifiedcommunications.techtarget.com/definition/real-time-communications

7 What is real time communications? http://www.realtimecommunicationsworld.com/topics/realtimecommunicationsworld/articles/376916-what-realtime-communications.htm

Resource allocation

- ✓ Operation license

Software

Regulation

- ✓ Technical standards
- ✓ Electric energy

Characteristics of a telecommunications system

- ✓ Interactions between elements for the same purpose
- ✓ Technical compatibility
- ✓ Use of the same procedures
- ✓ Response to control
- ✓ Harmonious operation

Information system

- ✓ Organized set of components for collecting, transmitting, storing and processing data to deliver information for action[8]
- ✓ Set of elements (personal, hardware, software ...) to acquire, process, store and communicate information[9]
- ✓ Examples of information system
  - o Transaction Processing System
  - o Decision Systems
  - o Knowledge system management

Functions of an information system

- ✓ Collection of information
- ✓ Storage of information
- ✓ Processing
- ✓ Communication or dissemination of information

Elements of an information system

- ✓ Source
- ✓ Encoder
- ✓ Transmitter
- ✓ Transmission channel
- ✓ Receiver
- ✓ Decoder
- ✓ Destination

Basic components of an information system

- ✓ Materials
- ✓ Softwares
- ✓ Procedures
- ✓ Data
- ✓ Equipment (Computer and Telecommunications)
- ✓ Human resources

Wired telecommunications systems

- ✓ Telecommunications system connecting users with cables
- ✓ System for exchanging information between two points using a tangible or physical medium

Main cables used

- ✓ Electrical cables (coaxial cables, twisted pairs)
- ✓ Optical cables

Examples of wired telecommunications systems

- ✓ Public Switched Telephone Network (PSTN)
- ✓ Cable TV system
- ✓ Internet Access by cable
- ✓ Optical fiber communications links

Wireless telecommunications systems

- ✓ Telecommunications systems connecting the user via electromagnetic waves
- ✓ System for exchanging information between two points without a physical medium

Types of wireless links used

- ✓ Broadcasting (by radio links for radio and television)
- ✓ Satellite links

8 Système d'information https://fr.wikipedia.org/wiki/Systeme_d'information

9 Système d'information de l'entreprise http://profs.vinci-melun.org/profs/adehors/CoursWeb2/Cours/Ch1/Ch1.php

- ✓ Microwave links
- ✓ Mobile communication systems
- ✓ Infrared links
- ✓ Wi-Fi (Wireless Fidelity)

Examples of wireless telecommunications systems

- ✓ Cellular systems
- ✓ Wireless LAN
- ✓ Satellite telecommunications systems
- ✓ Paging systems
- ✓ Bluetooth
- ✓ Ultra-wide band radios
- ✓ Zigbee Radios
- ✓ Satellite
- ✓ Free to air television
- ✓ Cordless telephone
- ✓ Wireless LAN, WIFI
- ✓ Wireless MAN, WIMAX
- ✓ Bluetooth
- ✓ Wireless Laser
- ✓ Microwave
- ✓ Global Positioning System (GPS)

Advantages of wireless telecommunications systems

- ✓ Mobility
  - o Link established by waves to guarantee freedom of movement
- ✓ Increased reliability
  - o Absence of physical infrastructure subject to breakage, vandalism, etc.
- ✓ Easy installation
  - o Fast connections with wireless networks
  - o Rapid Disaster Recovery
  - o Few equipment to restore
  - o Non deployment of cable
- ✓ Lower cost
  - o Decrease in expenses due to non-deployment of cable

Disadvantages of wireless telecommunications systems

- ✓ Radio interference
  - o Wireless links subject to radio interference from other wireless links
- ✓ Security
  - o Interception of radio waves transmitted by unauthorized users
  - o Confidentiality of information transmitted at risk[10]

Basic criteria in telecommunications

Reliability and fidelity

- ✓ *Reliability* **:** Permanence of service availability
  - o Unavoidable technical failures in telecommunications systems
  - o Reliability affected by the inability to respond at any time due to technical constraints
- ✓ *Fidelity:* Accurate restitution of information from source to destination thanks to the transparency of the telecommunications network
  - o Fidelity affected by distortions suffered during processing by the equipment and in the transmission channel
  - o Fidelity less than 100% given the imperfections of the technical means and disturbances.

RESSOURCES

Used in Telecommunications

Scarce and limited resources used by telecommunications

7 Resources used in the provision of electronic communications services

1. *Radio Frequency Spectrum*
2. *Numbering plan*

10 Wireless telecommunications – T.L. Singal https://books.google.ht/books/about/Wireless_Communications.html?id=cQJJzA8CCUUC&redir_esc=y

3- *Internet domain (domain extension)*
4- *High Points*
5- *Existing Infrastructures*
6- *IP Address (Addressing)*
7- *Orbital position (telecommunications satellite orbit)*

Resource 1: Radio Frequency Spectrum

- ✓ Set of frequencies used in telecommunications activities
- ✓ Set of defined frequencies for the provision of telecommunication services of all kinds
- ✓ Element of the public domain of the State, inalienable and nontransferable
- ✓ Portion of the electromagnetic spectrum used for telecommunications activities
- ✓ Portion of the electromagnetic spectrum (radio electricity, microwaves, infrared, ultraviolet, X-rays, gamma)
- ✓ Extent of the radio frequency spectrum: 9 KHz - 3000 GHz (defined by the ITU for telecommunications)
- ✓ Intangible ubiquitous resource in all electronic communications

Frequency

- ✓ Number of cycles per second of a phenomenon
- ✓ Number of periods per second of a vibratory movement
- ✓ Number of oscillations of a periodic phenomenon per unit of time
- ✓ Number of repetitions of a phenomenon during a given interval (periodic phenomenon)
- ✓ One of the characteristics of a signal being transmitted (Amplitude, Frequency and Phase)
- ✓ Intangible resource characterizing any signal
- ✓ Signal Uniqueness (A frequency associated with a signal)
- ✓ Unique identification of the presence of a signal in a given area
- ✓ Essential resource for the provision of wired and wireless telecommunications services

Frequency units

- ✓ Cycle per second (cycle/second, cps or c/s)
- ✓ Hertz (1 Cycle/second = 1 Hertz)

Multiple Hertz

- ✓ Kilohertz (KHz) = 1000 Hertz
- ✓ Megahertz (MHz) = 1000 KHz = 1000 000 Hertz (1 million Hertz or cycles per second)
- ✓ Gigahertz (GHz) = 1000 MHz = 1000 000 KHz (1 million KHz or Kilocycles per second = 1000 000 000 Hertz (1 billion Hertz)
- ✓ Terahertz (THz) = 1000 GHz = 1000 000 MHz (1 million MHz) = 1000 000 000 KHz (1 billion KHertz) = 1000 000 000 000 Hertz

State and use of the Radio Frequency Spectrum

- ✓ Element of the intangible heritage of the State
- ✓ State Monopoly on the Radio Frequency Spectrum
- ✓ Inexhaustible source of income for the State
- ✓ Means used by the State to control telecommunications activities in a country
- ✓ Establishment by the State of the management structure of this spectral resource

Uses of frequencies in electronic communications

- ✓ Element necessary for the deployment and operation of any telecommunications system (electronic communications)
- ✓ Essential element for access and use of telecommunication services (Transmitters and receivers tuned on the same frequency)
- ✓ Access to radio, television, telephony, Internet, thanks to the frequency
- ✓ Intangible infrastructure of telecommunications systems
- ✓ Frequency usage: Authorization or concession prior to the deployment of any telecommunications system
- ✓ A frequency band for each telecommunications service
- ✓ In telephony, a frequency in each direction of communication

- ✓ A frequency band for AM broadcasting: 530 - 1700 KHz
- ✓ 10 KHz: bandwidth assigned to an AM station
- ✓ A frequency band for FM broadcasting: 88 - 108 MHz
- ✓ 0.2 MHz or 200 KHz: bandwidth allocated to an FM radio station
- ✓ 54 - 60 MHz: frequency band of channel 2 (analog television)
- ✓ 76 - 82 MHz: frequency band of channel 5 (analog television)
- ✓ 800 MHz, 900 MHz, 1800 MHz and 1900 MHz: bands reserved for cellular telephony
- ✓ Diversity of uses of the radio frequency spectrum
  - o Maritime Transport (GMDSS), Recreational Boat
  - o Aeronautical transport: radio navigation
  - o Scientific Services: Meteorology, Radioastronomy
  - o Television and radiotelephony: terrestrial broacasting, satellite broadcasting
  - o Telecommunications: fixed service, mobile, worldwide satellite systems, local area networks
  - o Security networks: rescue, safety of life
  - o Defense: radars, etc.,
  - o Amateurs
  - o Standard frequencies and time signals[11]

Frequencies and transmission media

- ✓ Low frequencies: Wired communication over metallic cables (twisted pairs, coaxial cables, etc.)
  - o Copper: frequencies below 1MHz
- ✓ Medium frequencies: wireless communications (without a physical media)
  - o Wireless links: frequencies between MHz and tens of GHz
- ✓ High Frequencies: Wired Communications on Optical Fiber
  - o Optical fiber: frequencies greater than 30 GHz

Direct use of radio frequencies by the user

- ✓ Setting of a radio on a frequency
  - o XFM radio on 90.5 MHz frequency
- ✓ Setting of a television set on a channel (frequency band)
  - o Setting of the TV on the TV Excell channel (channel 5: 72 -78 MHz)
- ✓ Phone call (one frequency for each direction of communication)

Radio frequency management

Set of technical and administrative procedures for the use of frequencies without harmful interference

- ✓ Allotment
- ✓ Assignment
- ✓ Notification
- ✓ Coordination
- ✓ Control
- ✓ Recording[12]

Players involved in the management of the radio frequency spectrum

- ✓ International Telecommunication Union (principles and methods of management)
- ✓ National regulatory bodies (frequency spectrum management)
- ✓ Network operators (use of frequency bands)
- ✓ Equipment manufacturers (design and development of equipment for frequency use)

Radio frequency spectrum management functions

Main functions

Allocation (of a frequency band)

- ✓ Inclusion in the frequency band allocation table of a specified frequency band for use by one or more *terrestrial or space* radiocommunication

11 Gestion du spectre des fréquences - Outils d'aide à l'assignation des fréquences – https://www.itu.int/ITU-D/tech/events/2009/RDF_ARB/Presentations/Session7/RDF09_ARB_Presentation_IBChaabane.pdf

12 https://www.itu.int/en/itunews/Documents/2015_ITUNews05-fr.pdf

services or by the *radio astronomy service* under specified conditions[13]

- ✓ Distribution of the usable part of the radio spectrum in small bands to be allocated exclusively or shared to specific radiocommunication services

Assignment (of a frequency or radio channel)

- ✓ Authorization given by an administration for use by a *radio electric* station of a specified frequency or radio channel under specified conditions[14]

Allotment (of a frequency or a radio channel)

- ✓ Inclusion of a given channel in a plan adopted by a competent conference, for use by one or more administrations for a terrestrial or space *radiocommunication service*, in one or more specified countries or geographical areas and under specified conditions[15]
- ✓ Frequency allocation: Reservation or allocation of frequencies to telecommunications services
- ✓ Frequency Assignment: Authorization granted to a radio station to use a given frequency
- ✓ Frequency Allotment: Reservation or allocation of frequencies to geographical areas or countries

13 Le partage des fréquences – ITU – https://www.itu.int/newsarchive/press/WRC97/Sharing-the-spectrum-fr.html

14 Le partage des fréquences – ITU https://www.itu.int/newsarchive/press/WRC97/Sharing-the-spectrum-fr.html

15 Le partage des fréquences – ITU https://www.itu.int/newsarchive/press/WRC97/Sharing-the-spectrum-fr.html

Different radio frequency bands

Table showing the frequency bands, their designations, their wavelengths and their uses in telecommunications

Band	Designation	Wave length	Use
3 - 30 KHz	Very low frequencies	100 Km - 10 Km	Underwater Communications, Port-Boat Communications
30 - 300 KHz	Low frequencies	10 Km - 1 Km	Great Waves - Sound Broadcasting
300 - 3000 KHz	Medium frequencies	1 Km - 100 m	Average waves (AM) Distress frequency (500 KHz) Weather Signals
3 - 30 MHz	High frequencies	100m - 10m	Shortwave broadcasting Amateur Radio Aviation
30 - 300 MHz	Very high frequencies (VHF)	10 m - 1m	FM band - Hertzian TV bands - professional radio networks - CB - private networks ...
300 - 3000 MHz	Ultra-High Frequency (UHF)	1m - 10 cm	TV bands - GSM / DCS / UMTS telephony - Satellite communication Microwave transmission
3 - 30 GHz	Super High Frequencies (SHF)	10 cm - 1cm	Satellites TV Wireless links Radar
30 - 300 GHz	Extremely High Frequencies	1cm - 1mm	Radars Satellites
300 - 3000 GHz	Tremendously High Frequencies	1 mm - 1 micro	
3 – 30 THz		cmm	
30 – 300 THz		µm	
300 – 3000 THz		dµm	

Relationship between frequency and velocity

- ✓ Propagation velocity of waves in free space: speed of light (C = 300 000 km / s)
  - o C = speed of light (speed), F = frequency and λ = wavelength
  - o λ = c: f
- ✓ λ in meters, c in meters per second (m/s) and f in Hertz
- ✓ f = 100 MHz, λ = 3 meters

Free frequencies

Frequencies used worldwide without authorization

- ✓ 2.4 GHz band
- ✓ 5.8 GHz band
- ✓ 24 GHz band

Stakes of the radio frequency spectrum

- ✓ Limited and rare resources
- ✓ Essential Resources for the deployment and operation of Telecommunication networks

Resource 2: Telephone numbering plan

- ✓ Numbering system used in telecommunications to assign telephone numbers according to the origin of the telephone call[16]
- ✓ Rare resource consisting of all telephone numbers allowing to identify fixed or mobile endpoints of telephone networks and services, route calls and access internal network resources
- ✓ Numbering assignment for the provision of telephone Service

Telephone numbers

- ✓ Sequence of ordered numbers uniquely indicating the endpoint of the public network
- ✓ Unique identification of a subscriber in a telephone network
- ✓ Number containing the information needed to route calls to the endpoint
- ✓ Medium used by the telecommunications system to serve subscribers
- ✓ User-friendliness between numbers and telecommunications systems (Identification of users by numbers, not by names)
- ✓ Worldwide uniqueness of the telephone number thanks to the country code
- ✓ State resources assigned to telephony operators for the provision of telephone service
- ✓ Assignment of a telephone number to each telephone subscriber by the telephone operator for accessing and using the telephone service

Format of a telephone number

- ✓ + Country Code + Area Code / Mobile Code + Subscriber Number
- ✓ *Subscriber Number*: Last 4 of the telephone number (subscriber identification)
- ✓ In Haiti : 509 + 3700 + xxxx (xxxx = number of the subscriber)
- ✓ *Format of a fixed number in Switzerland* : 41 + 2 22 234+ xxxx (xxxx = number of the subscriber)
- ✓ *Format of a mobile number in Switzerland* : 41 + 79 567+ xxxx (last 4 digits of the number = subscriber number)
- ✓ *PLUS sign (+)*: Reminder to the user for the addition of the international prefix
- ✓ *Length of a telephone number*: 15 digits according to ITU recommendation E.164

Types of telephone numbers

- ✓ *Geographical* numbers: numbers indicating zones of a given country (application in fixed telephony only)
- ✓ *Mobile numbers*: Numbers assigned to cellular telephony for use in a given country (telephone numbers assigned to cellular subscribers regardless of residential area)
- ✓ *Non-geographical numbers*: National telephone numbering plan telephone numbers assigned to a specific area and intended to receive only incoming telephone calls (designed for special services such as toll-free telephone number: 1800 in the United States of America, 0800 in France)
- ✓ *Short numbers*: Telephone numbers consisting of 3 or 4 digits

16 Plan de numération
https://fr.wikipedia.org/wiki/Plan_de_numérotation

Uses of short telephone numbers

2 possible uses

- ✓ Emergency services (police, ambulance, fire ... ...)
  - o 911 in the United States of America
- ✓ Customer services (telephony operators, commercial enterprises or institutions)
  - o Voice communication
  - o Text message exchange

Tool-Free numbers

- ✓ Telephone number allowing to call an organization or institution without charge from the caller (cost of the call to be borne by the called party)
- ✓ Examples of Tool-Free numbers
  - o 0800 xxx xxx in France
  - o 1800 xxx xxxx in the United States of America

Guidelines for the telephone numbering plan

Possible orientations

Orientation 1: Orientation based on area

- ✓ Assignment of an area code (in fixed telephony, identifying a zone or town in a country by a dialing code)

Orientation 2: Orientation based on Telephone Operators

- ✓ Assignment of a telephone code to each telephone operator as a means of identification (numbers starting with 2 for the telephone operator X, numbers starting with 3 for the telephone operator Y, etc.)

Orientation 3: Orientation based on service

- ✓ Assignment of a telephone area code to each category of telephone service as a means of identification
- ✓ Examples:
  - o 1 for special services: Emergency and assistance
  - o 2 for fixed telephone service
  - o 3 and 4 for cellular telephone service
  - o 8 for value-added services and green numbers, and 9 for IP telephony (VoIP)

Country codes

- ✓ Means of identification of a country and of routing of calls and messages
- ✓ Country code: 2 or 3 digits
- ✓ Country code with 2 digits: 33 (France), 41 (Switzerland), 86 (China), 27 (South Africa)
- ✓ Country code with 3 digits: 509 (Haiti), 250 (Rwanda), 256 (Uganda), 351 (Portugal), 420 (Czech Republic)

International telephone number prefix

- ✓ Means of indicating to the telecommunications network the international destination of the telephone call
- ✓ Means of distinguishing international calls from national calls
- ✓ International prefix: 2 or 3 digits
- ✓ International prefix with 2 digits: 00
- ✓ International prefix with 3 digits: 001 or 011

Worldwide numbering plan

- ✓ 9 numbering zones defined around the world
- ✓ Zone 1 countries (United States, Canada and some Caribbean countries)

4 particularities in zone 1

Particularity 1

- ✓ **1**: Country code for the United States of America and Canada
  - o USA, Canada: use of area code (not a traditional country code)
  - o + 1 area code (3 digits) + telephone number (7 digits)
  - o + 1 305 + xxx xxxx (Miami, Florida)
  - o +1 301 + xxx xxxx (Maryland)
  - o +1 514 + xxx xxxx (Quebec)

Particularity 2

- ✓ Zone 1 countries with different codes for fixed and mobile numbers
- ✓ Example: Dominican Republic
  - o Country codes for fixed numbers: 809 and 829
  - o Country code for mobile numbers: 849

Particularity 3

- ✓ Digit 1: Reminder to users about the zone of destination (zone 1) of the call or SMS / MMS (Zone 1)

Particularity 4

- ✓ Use of digits 2, 3, 4, 5, 6, 7, 8 and 9 as the first digits in codes of zone 1, and differentiation established by digit 1

Countries of the world (9 zones) and their telephone codes

Zone 1

Codes	Countries	Codes	Countries
+1	United States of America	+1 (340)	Virgin Islands of the United States
+1	Canada	+1 (345)	Cayman Islands
+1 (242)	Bahamas	+1 (441)	Bermuda
+1 (246)	Barbados	+1 (473)	Cariacou (Grenadines)
+1 (264)	Anguilla	+1 (649)	Turkish and Caicos Islands
+1 (268)	Antigua and Barbados	+1 (664)	Montserrat
+1 (284)	British Virgin Island	+ 1 (670)	Northern Mariana Islands
+1 (671)	Guam	+1 (684)	American Samoa
+1 (721)	Saint Maarten	+1 (758)	St. Lucia
+1 (767)	Dominica	+1 (784)	Saint Vincent and the Grenadines
+1 (787)	Porto Rico	+1 (809)	Dominican Republic
+1 (829)	Dominican Republic	+1 (849)	Dominican Republic
+1 (868)	Trinidad and Tobago	+1 (869)	St. Kitts and Nevis St. Kitts and Nevis
+1 (876)	Jamaica	+1 (939)	Puerto Rico

Zone 2 (African countries and countries from other continents)

Codes	Countries	Codes	Countries
+20	Egypt	+211	South Sudan
+212	Morocco	+213	Algeria
+216	Tunisia	+218	Libya
+220	Gambia	+221	Senegal
+222	Mauritania	+223	Mali
+224	Guinea	+225	Ivory Coast
+226	Burkina Faso	+227	Niger
+228	Togolese Republic	+229	Benin
+230	Mauritius	+231	Liberia
+232	Sierra Leone	+233	Ghana
+234	Nigeria	+235	Tchad

+236	Central African Republic	+237	Cameroon
+238	Cape Verde	+239	Sao Tome and Principe
+240	Equatorial Guinea	+241	Gabon
+242	Congo Brazzaville)	+243	Congo (Kinshasa)
+244	Angola	+245	Guinea Bissau
+246	Diego Garcia	+247	Ascension
+248	Seychelles	+249	Sudan
+250	Rwanda	+251	Ethiopia
+252	Somalia	+253	Djibouti
+254	Kenya	+255	Tanzania
+256	Uganda	+257	Burundi
+258	Mozambique	+260	Zambia
+261	Madagascar	+262	Mayotte and Reunion
+263	Zimbabwe	+264	Namibia
+265	Malawi	+266	Lesotho
+267	Botswana	+268	Swaziland
+269	Comoros	+27	South Africa
+291	St. Helena and Tristan da Cunha	+292	Eritrea
+297	Aruba	+298	Faroe Islands
+299	Greenland		

Zone 3 (European countries)

Code beginning with the number 3

Codes	Countries	Codes	Countries
+30	Greece	+31	Netherlands
+32	Belgium	+33	France
+34	Spain	+350	Gibraltar
+351	Portugal	+352	Luxembourg
+353	Ireland	+354	Iceland
+355	Albania	+356	Malta
+357	Cyprus	+358	Finland
+358	Bulgaria	+36	Hungary
+370	Lithuania	+371	Latvia
+372	Estonia	+373	Moldova

+374	Armenia	+375	Belarus
+376	Andorra	+377	Monaco
+378	San Marino	+379	Vatican City
+380	Ukraine	+381	Serbia
+382	Montenegro	+385	Croatia
+386	Slovenia	+387	Bosnia and Herzegovina
+389	Macedonia	+39	Italy

Zone 4 (other European countries)

Codes	Countries	Codes	Countries
+40	Romania	+41	Switzerland
+420	Czech Republic	+421	Slovakia
+423	Liechtenstein	+43	Austria
+44	UK	+45	Denmark
+46	Sweden	+47	Norway
+48	Poland	+49	Germany

Zone 5 (South America and Latin America)

Code beginning with the number 5

Codes	Countries	Codes	Countries
+500	Falkland Islands	+501	Belize
+502	Guatemala	+503	El Salvador
+504	Honduras	+505	Nicaragua
+506	Costa Rica	+507	Panama
+508	St Pierre and Miquelon	+509	Haiti
+51	Peru	+52	Mexico
+53	Cuba	+54	Argentina
+55	Brazil	+56	Chile
+57	Colombia	+58	Venezuela
+590	Guadeloupe	+591	Bolivia
+592	Guyana	+593	Ecuador
+594	French Guiana	+595	Paraguay
+596	Martinique	+597	Suriname
+598	Uruguay	+599	Bonaire, Saba, Curacao, Saint Eustatius

Zone 6 (countries of Oceania and South Pacific)

Code beginning with the number 6

Codes	Countries	Codes	Countries
+60	Malaysia	+61	Australia
+62	Indonesia	+63	Philippines
+64	New Zealand	+65	Singapore
+66	Thailand	+670	East Timor
+672	Foreign Territories of Italy	+673	Brunei Darussalam
+674	Nauru	+675	Papua New Guinea
+676	Tonga	+677	Solomon Islands
+678	Vanuatu	+679	Fiji Islands
+680	Palau	+681	Wallis and Futuna
+682	Cook Islands	+683	Niue
+685	Samoa	+686	Gilbert Islands (Kiribati)
+687	New Caledonia	+688	Tuvalu
+689	French Polynesia	+690	Tokelau
+691	Micronesia	+692	Marhalls Islands

Zone 7 (Russia and neighboring countries (former USSR))

Code beginning with the number 7

Codes	Countries	Codes	Countries
+7	Kazakhstan	+7	Russia

NB: Sharing of county code +7 between Russia and Kazakhstan

Possible assignment in the future of codes +990, +997 and +999 to Kazakhstan

Zone 8 (East Asia and special services)

Code beginning with the number 8

Codes	Countries	Codes	Countries
+81	Japan	+82	South Korea
+84	Vietnam	+851	North Korea
+852	Hong Kong	+853	Macao
+855	Cambodia	+856	Laos
+86	China	+880	Bangladesh
+886	Taiwan		

Special Services Operating Number 8

Special services numbers

Codes	Services	Codes	Services
+800	International Freephone	+808	Shared services
+870	Inmarsat «SNAC» Service	+871	Inmarsat (Atlantic East)
+872	Inmarsat (Pacific)	+873	Inmarsat (Indian)
+874	Inmarsat (Atlantic West)	+878	Universal personal telecommunications
+881	Global mobile satellite system	+882	International Networks
+883	International Networks	+888	Telecommunications for emergency relief

Inmarsat (International Maritime Satellite Organization): a company in the telecommunications sector specialized in the provision of mobile-satellite service

Zone 9 (countries from West and South Asia and the Middle East)

Code beginning with the number 9

Codes	Countries	Codes	Countries
+90	Turkey	+91	India
+92	Pakistan	+93	Afghanistan
+94	Sri Lanka	+95	Burma (Myanmar)
+960	Maldives	+961	Lebanon
+962	Jordan	+963	Syria
+964	Iraq	+965	Kuwait
+966	Saudi Arabia	+967	Yemen
+968	Oman	+970	Palestine
+971	United Arab Emirates	+972	Israel
+973	Bahrain	+974	Qatar
+975	Bhutan	+976	Mongolia
+977	Nepal	+98	Iran
+992	Tajikistan	+993	Turkmenistan
+994	Azerbaijan	+995	Georgia
+996	Republic of Kyrgyzstan	+998	Uzbekistan

Source : World Telephone Numbering Guide- www.wtng.info

Dialing of an international telephone number (international call)

- ✓ International prefix + country code + area code or mobile code + subscriber number
- ✓ Call to any country: 3 possible cases

- o 00 + country code + telephone number
- o 001 + country code + telephone number
- o 011 + country code + telephone number

Zone 1 to zones (2, 3, 4, 5, 6, 7, 8, 9)

- ✓ Zone 1 to the rest of the world (zones: 2, 3, 4, 5, 6, 7, 8, 9): 011 + country code + area code / mobile code + subscriber number
- ✓ Examples: A call from the United States to Switzerland: 011+ 41+ remainders of the phone number
- ✓ A call from Canada to Gabon: 011+ 241+ remainders of the phone number
- ✓ A call from Jamaica to Japan: 011+ 81+ remainders of the phone number

Zones (2, 3, 4, 5, 6, 7, 8, 9) to zone 1

- ✓ Rest of the world (Zone 2 - Zone 9) to zone 1: 001 + country code + area code / mobile code + subscriber number
- ✓ Examples: A call from France to the United States of America
  - o 001 + 305 .... (for a call to Miami, Florida)
  - o A call from Mali to Puerto Rico: 001 + 787 + remainders of the telephone number
  - o A call from Indonesia to the Bahamas: 011 + 242 + remainders of the telephone number
  - o A call from Haiti to the Dominican Republic: 001+ 809 + remainder of the telephone number or 001+ 829 + remainder of the telephone number or 001+ 849 + remainder of the telephone number

Zones (2, 3, 4, 5, 6, 7, 8, 9) to countries of zones (2, 3, 4, 5, 6, 7, 8, 9)

- ✓ Countries of zones (2, 3, 4, 5, 6, 7, 8, 9) to country of zones (2, 3, 4, 5, 6, 7, 8, 9): 00 + country code + subscriber number
- ✓ Examples
  - o A call from Germany (zone 4) to India (zone 9): 00 + 91+ Subscriber number
  - o A call from Malaysia (zone 6) to South Korea (zone 8): 00 + 82 + Subscriber number
  - o A call from Russia (zone 7) to Panama (zone 5): 00 + 507 + Subscriber number
  - o A call from Burundi (zone 2) to Sweden (zone 4): 00 + 46 + Subscriber number

Format of a satellite phone number

- ✓ Satellite prefix + satellite telephone number (8 or 9 digits)
- ✓ Examples
  - o 88+ 8 digits (Iridium)
  - o 870 + 9 digits (Inmarsat)
  - o 882 16 + 8 digits (Thuraya)
  - o *8818 + 8 digits (*Globalstar)
  - o 8819 + 8 digits (Globalstar)

Call from a terrestrial telephone (fixed or mobile) from zone 1 to a satellite phone (whatever the geographical position of the satellite)

011+ satellite prefix + satellite telephone number

Example: A call from the United States of America to satellite telephones

- o 011 + 88 16 + satellite telephone number (Iridium)
- o 011 + 870 + satellite telephone number (Inmarsat)
- o 011 + 882 16 + satellite telephone number (Thuraya)

Call from a terrestrial telephone (fixed or mobile) from zones (2 to 9) to a satellite telephone (whatever the geographical position of the satellite)

00 + satellite prefix + satellite telephone number

Example: A call from any country, except zone 1

- o 00 + 88 16 + satellite telephone number (Iridium)
- o 00 + 870 + satellite telephone number (Inmarsat)
- o 00 + 882 18 + satellite telephone number (Thuraya)

Call from a satellite telephone to a terrestrial telephone (fixed or mobile) in zone 1 (whatever the geographical position of the satellite)

- 001+ country code (area code for USA) + telephone number
- 001+ 305 + telephone number (A call from a satellite telephone to a phone based in Miami, Florida, USA)
- 001+ 876 + telephone number (A call from a satellite phone to a phone based in Jamaica)

Call from a satellite telephone to a terrestrial telephone (fixed or mobile) in zones (2 to 9) (regardless of the geographical position of the satellite)

- 00 + country code + telephone number
- 00 +33 + telephone number (to France)
- 00 +509 + telephone number (to Haiti)
- 00 + 44 + telephone number (to the United Kingdom)
- 00 + 86 + telephone number (to China)

Call from a satellite telephone to another satellite telephone (regardless of the geographical position of the satellite)

A call from an Iridium satellite phone to an Iridium satellite telephone

- 00 + satellite prefix +telephone number
- 00 + 88 16 + telephone number

Resource 3: Internet domains (domain extensions)

- ✓ Identification of a country on the Internet
- ✓ Country Internet Domain: Country Internet Code or Country Extension

Purpose of the domain name

- ✓ Facilitation of access to the website for the user

Example: 17. 123. 6. 239 (address corresponding to the domain name of the website "www.ht.com")

Country domains (domain name extensions)

- ✓ Code consisting of **2 letters** to identify a country on the Internet (similar to a telephone country code)
- ✓ Suffix at the end of the address of a website (www. Primature.ht, www. rfi.fr)
- ✓ Country extension: code attached to websites and e-mail addresses of a given country
- ✓ National extension corresponding to the ISO code 3166
- ✓ First level extension associated with a country
- ✓ Means of identification of the citizens, companies and institutions of each country
- ✓ Country extension: referencing of Internet addresses and easy access to national websites
- ✓ Examples:
    - fr (France), ch (Switzerland), ht (Haiti), uk (United Kingdom), us (United States of America), ca (Canada), ci (Ivory Coast), ga (Gabon), jp (Japan), cn (China), au (Australia)

Other domains

Domain names used for other fields of activity

- ✓ .com (commercial organizations)
- ✓ .org (non-profit organization)
- ✓ .net (large networks)
- ✓ .gov (governmental organization)
- ✓ .edu (education)
- ✓ .mil (Military Institution)
- ✓ .int (international institutions)

Resource 4: High Points

- ✓ Very high geographical space used for the transmission of electromagnetic waves
- ✓ Geographic space used for the implementation of transmission sites (transmission and reception)

2 fundamental conditions of a high point

- ✓ Height: Ability to have an altitude with high visibility
- ✓ Accessibility: ability to establish an access route to install, maintain and provide electrical power to transmission equipment

Resource 5: Existing infrastructures

- ✓ Infrastructure already deployed and likely to be used by a telecommunications operator
- ✓ Examples
    - Power grid for the use of power line technology (powerline communication)

- o Telecommunications tower (for the installation of radio equipment: transceiver, antennas, etc.)
- o Cable transmission network

Resource 6: IP address

- ✓ IP: Internet Protocol
- ✓ Address identifying a terminal connected to the Internet
- ✓ Quantity of IP = Quantity of terminals connected to the Internet
- ✓ IP address: Groups of numbers separated by dots
- ✓ IPv4 example (IP version 4): 17.223.125. 86
- ✓ Each group: a number between 0 and 255
- ✓ IPv4: 4,294,967,296 IP addresses
- ✓ IPv4: 32-bit IP address system
- ✓ IPv4: address exhaustion
- ✓ Transition to IPV6 for more IP addresses
- ✓ IP V6 format: 8 groups of 4 hexadecimal digits
- ✓ IPV6 example: 2001: 0660: 7401: 0200: 0000: 0000: 0edf: bdd7
- ✓ IPv6: 128-bit IP address system
- ✓ IPV6: 3.4×10^{38} IP addresses
- ✓ IPV6: 3.4×10^{38}terminals connected to the Internet

IP address functions

- ✓ Unique identification of each computer or other terminals connected to the Internet
- ✓ Protocol necessary for any exchange via the Internet (communication between the connected machine and other connected computers)
- ✓ Transmission and reception of data
- ✓ Addressing of packets transmitted over the Internet
- ✓ Assembly and disassembly of packets during transmission
- ✓ Delivery of packets

Resource 7: Orbital positions

- ✓ Trajectory intended to receive a telecommunications satellite (geostationary or not)
- ✓ Position required to place an artificial satellite in space
- ✓ Geostationary orbit or geosynchronous: 35876 km to the ground
- ✓ Low orbits: 180 - 400 Km (LEO), 400 - 1000 Km (MEO), 400 - 900 Km (Heli synchronous orbit)
- ✓ Orbit position: determination of radio signal coverage
- ✓ Limited number of available orbits
- ✓ Applications to be submitted to ITU for orbital position allocation and concession for all jurisdictions to be served)

Resources Used to establish an Electronic Communication

Use of multiple resources to establish an electronic communication
Number of resources used related to the type of service provided

- ✓ Telecommunications resources (transmitters, receivers, terminals, antennas, transmission lines, switches, etc.)
- ✓ Computing Resources (Computer hardware, software, database, servers, routers, gateways, etc.)
- ✓ Energy Resources (City Current or town current, Generators, Backup System, etc.)
- ✓ Logistics for Equipment (Buildings, Air Conditioning)

CHAPTER 2

TELECOMMUNICATIONS AND ICTs ENVIRONMENT

MAIN VARIABLES OF THE TELECOMMUNICATIONS/ICTs ENVIRONMENT

- ✓ Technological Variable
  - o Technologies: Basis of telecommunications services
  - o Technologies: Pillar and lever of telecommunication development
  - o Huge investments to be made to use the latest technologies
- ✓ Legal variable
  - o Legal framework of the sector
  - o Legal framework for the provision of services
  - o Impacts of legal decisions on the sector
- ✓ Economic variable
  - o Second world economy
  - o Definition of business models for the telecommunications sector
  - o Reflection of operations expenses on the cost of services
  - o Cost of Infrastructure Deployment in Rural Areas
- ✓ Social variable
  - o Limited access in rural areas
  - o Requirement of package offer for multimedia communications
  - o Rush towards online transactions
- ✓ Political variable
  - o Control instrument for the State
  - o Regulation orientation from the State
- ✓ Environmental variable
  - o Impact of environmental factors on service provision
  - o Adaptation of technologies to climate change.

Telecommunications Activities

- ✓ Establishment and/or operation of electronic communications networks and services
- ✓ Manufacturing, import and export of telecommunications equipment
- ✓ Advertising and sale of telecommunications equipment
- ✓ Use and installation of telecommunications equipment[17]

PLAYERS IN THE TELECOMMUNICATIONS SECTOR

- ✓ Consumers
  - o Users or consumers of telecommunications services
  - o Examples: telephone subscribers, listeners, viewers, Internet users. Companies and institutions
- ✓ Broadcasting Operators
  - o Operators licensed for the provision of broadcasting services
  - o Examples: Sound and television broadcasting stations
- ✓ Cable television operators
  - o Companies licensed to provide cable television service
  - o Examples: Telenet, Numericable, Comcast, Videotron, Rogers Cable, Cox Communication, Time Warner Cable, Virgin Media
- ✓ Telephony Operators
  - o Operators licensed for the provision of telephony services
  - o Examples: T-Mobile, Vodafone, British Telecom, Deutsche Telecom, France Telecom, NTT, Telefonica, Verizon, British Telecom, Digicel
- ✓ Network Service Provider

17 Code Des Télécommunications http://www.juristeconsult.net/ministere_justice/?code=code_20160525201037

- Enterprise providing Service Providers with Aces Network and Bandwidth
- Examples: Telephone network infrastructure deployed to serve telephone service providers
- Broadband Data Network for Internet Service Providers

✓ Mobile Virtual network operator (MVNO)

- Operator licensed to provide mobile telephony service supported by a network operator's infrastructure
- Examples: Lebara, Lycamobile, Mobistar, Ortel Mobile

✓ Internet Service Providers

- Companies licensed to provide access to the Internet
- Examples: Communications Cox, Odynet, Teleconnect, Free, Fastweb

✓ Telecommunications Satellite Operators

- Companies providing satellite link services
- Examples: Intelsat, Arabsat, Eutelsat

✓ Operators of satellite telecommunications services

- Enterprises licensed for the provision of satellite telecommunications services
- Examples: Iridium, Globalstar, Inmarsat, Thuraya , ACeS

✓ Satellite TV Operators

- Companies licensed for the provision of satellite television services
- Examples: Canalsat, DirecTV, Dish Network, BSkyB

✓ Earth station operators

- Companies specialized in the establishment of connections with telecommunications satellites

✓ Optical cable operators

- Companies licensed for Optical Cable Deployment for the Transport of Telecommunications Services
- Examples: Telcité, Ariège Telecom, Inolia

✓ Transport Operators (Carriers)

- Company specialized in the transport of telephone traffic between two telephone operators
- Examples; Verizon, T-Mobile, AT & T, Sprint

✓ Equipment manufacturers

- Companies specialized in the manufacture of telecommunications equipment of all kinds (transmission equipment, switching and terminals)
- Examples: Alcatel -Lucent, Motorola, Nokia, LG, Sony, Fujitsu, Ericsson, Nortel, ZTE, Samsung, Apple (terminal equipment)
- Examples: Cisco, Lucent, Motorola (communication equipment)
- Examples: **Huawei, ZTE,** Ericsson(transmission equipment)
- Examples: Cisco., Huawei, Nokia, Alcatel-Lucent, ZTE (switching equipment)
- Examples: Cisco, Nortel, 3com, Siemens Nokia, Alcatel Lucent, Ericsson, Huawei (network equipment)
- Examples: Motorola, Qualcomm, Sony, NEC, COMCAST, Boeing, Lockheed-Martin, EADS-Astrium, Thalès Alena Space, Loral (wireless and satellite communication equipment)
- Examples: Hewlett - Packard, Cisco (computer network equipment)

✓ Standardization Agencies

- Institutions specialized in the design and definition of standards for the telecommunications sector
- Examples: IEEE, ETSI, ISO, ITU

✓ Regulators

- Authorities responsible for regulating the telecommunications sector
- Examples: Federal Communication Commission (USA), ARCEP (France), CONATEL (Hai-

ti), INDOTEL (Dominican Republic)

✓ Software and Application Developers (for Service Providers and the General Public)

  - Companies designing and developing software and specialized applications for the market
  - Examples: Microsoft, Apple, IBM, Google

✓ Content Editors

  - Company specialized in the edition of multimedia content for broadcasting by television operators of all kinds
  - Examples: France Television, TF1, Canal +, M6, Radio France

✓ Network Service Providers

  - Company offering a high-speed data transmission network to Internet service providers or other businesses that need a fast link between their network and the Internet[18]
  - Examples: (PSTN, mobile operator, cable operator (ATT, Comcast, Verizon, DirectTV)

✓ Semiconductor Manufacturers (Resistors, Capacities, Diodes, Transistors)

  - Company specialized in the manufacturing of semiconductors
  - Examples: Texas Instruments, Qualcomm Broadcom, STMicroelectronics

✓ Operating System Providers

  - Companies specialized in the design and development of operating systems for computers, tablets, cellular telephones
  - Microsoft Corporation, Intel Corporation, IBM, Oracle, Google

✓ Cloud Computing Provider

  - Company specializing in the provision of cloud computing service (remote storage and computing service)
  - Google, Oracle, IBM, HP, Amazon, Microsoft

✓ Utility Service Providers

  - Company specialized in providing real-time services
  - Examples: AOL and Microsoft Instant Messaging
  - Videoconferencing and videoconferencing

✓ Consumer Electronics Suppliers

  - Manufacturer of electronic equipment for use by the general public
  - Examples: Motorola, LG, Samsung, Nokia, ZTE, Apple (Handheld Devices)
  - Examples: Polycom (videoconferencing terminals)
  - Examples: Microsoft and Sony (video game terminals)
  - Examples: Acer, Apple, Dell, HP or Toshiba (micro-computers)
  - Examples: Panasonic, Sharp or Sony (MP3 players, digital cameras and TV receivers)

✓ Online Application Provider (Application Service Provider or Hosted Application Provider)

  - Company providing software or computer services to customers through a network (Internet in general)
  - Examples: Hotmail, Yahoo Mail, Gmail, Google Spreadsheet, Google Docs,Free Online Logo Makers[19]

✓ Service Providers (Service Platforms and Intermediaries)

  - Company providing services through operator networks
  - Examples: Search engines (Google or Yahoo)
  - Service suites (Googlemaps, gmail, Flicker ...)
  - Online sales (Amazon, eBay, Expedia, private sales.com ...)
  - Social networks (Facebook, Twitter, Instagram)[20]

18 Bibliothèque virtuelle – https://www.oqlf.gouv.qc.ca/ressources/bibliotheque/dictionnaires/Internet/fiches/8359313.html

19 Fournisseur de services d'applications – https://fr.wikipedia.org/wiki/Fournisseur_de_services_d'applications

20 Les opérateurs de réseaux dans l'économie numérique – Lignes

- ✓ Digital content producers
  - o Company specialized in the production of contents
  - o Examples: textual information, images, music and movies, self-produced content
  - o Examples of producers: Time Warner, Walt Disney
  - o Individual producers: social network users, blogs

Telecommunications Sector Organization and Sector Functions

3 *Levels of organization*

1- Political Mission

- ✓ Policy defined by the government and parliament
- ✓ Governance of the telecommunications sector
- ✓ Definition of the vision and strategic directions of the sector
- ✓ Proposed law on telecommunications
- ✓ Approval of Telecom Law

2- Regulatory mission

- ✓ Regulation by a national telecommunications regulatory authority

3- Operation mission

- ✓ Operations based on direct services provided by the providers (Telecoms Operators, Broadcasting Operators, Internet Service Providers, Satellite Operators, etc.)

Telecommunications Policy

- ✓ Roles of Telecoms / ICTs in the economy and society
- ✓ Balancing between public and private interest through local, regional, national and international regulation of a growing variety of communication technologies
- ✓ Dynamic political process of distribution of costs and benefits across all these telecommunications sub-sectors

Objectives of telecommunications policy

- ✓ Regulation
- ✓ Efficient operation of the sector
- ✓ Telecommunications Sector Development
- ✓ Creation of an environment conducive for competition in the sector

Areas of intervention of the telecommunications policy

- ✓ Management of State resources used in telecommunications
- ✓ Economic and technical regulation
- ✓ International Telecommunications
- ✓ Protection of the interests of the State, operators and consumers
- ✓ Governance of the Telecoms/ICTs Sector
- ✓ Entrepreneurship in the Telecoms/ICTs sector
- ✓ Innovation in the Telecoms/ICTs sector

Elements of telecommunications policy

Policy Development

Telecoms/ICTs Laws

Recommendations

Telecommunications Services Regulations

Stakeholders of telecommunications policy

- ✓ Government
- ✓ Parliament
- ✓ Ministry of ICTs
- ✓ Regulatory Agency
- ✓ Operators and service providers
- ✓ Civil society and consumer associations

Strategies of telecommunications policy

- ✓ Industry coaching Policy
  - o Promotion of competition and innovation
  - o Promotion of universal and affordable access to telecommunications services
  - o Policy development
  - o Policy enforcement
- ✓ Regulatory coaching Policy

- o Adaptation of the regulatory framework to technological progress

✓ International Telecommunications Policy

- o Negotiation of bilateral and multilateral agreements[21]

Divisions of the telecommunications sector

✓ Telecommunications (telephony, radiotelephony, telegraphy, etc.)

✓ Sound and television broadcasting

✓ Data Transmission (Data Transmission and Internet)

State Public Power and Telecommunications

✓ Telecommunications: Element of the public power of the State

✓ Telecommunications: Means used by the state to ensure the security of life and the national territory

✓ Roles of the state in the provision of telecommunications services to citizens (such as paid basic services : Drinking water, energy, transport, postal services, etc.)

✓ Control of all telecommunications activities on the national territory

✓ Control of radio signals beyond the country's borders (control of the level of radiated power in neighboring territories for the reduction of the risk of radio interference with other telecommunications systems in operation in the neighboring country and the prevention of illegal trade)

✓ Regulation of International Electronic Communications: International telecommunication (telephony) traffic, access to the Internet through an international connection, broadcasting of radio and television programs across borders

✓ Authorization necessary for the exercise of any telecommunications activity

✓ Management of State resources used for the provision of telecommunications services

✓ Use of the telecommunications sector by the State (State: first telecommunications operator)

✓ Use of telecommunications signals in the territory of a country (Control of telecommunications activities in the territory)

Monopoly on the Telecommunications Sector

✓ De facto monopoly of the State on the sector of Telecommunications

✓ Monopoly granted to the State by the Telecommunications Sector Law

✓ Historical Telecommunications Operator

✓ Example: Historical Operator: State property

- o First telecommunications operator in a country

Telecommunications Ecosystem

Different elements of the ecosystem

1.- Governance of telecommunications

✓ Political Mission

✓ Regulation

2.- Telecommunications Indicators

✓ Services

✓ Broadcasting (sound and television)

✓ Telephony (fixed telephony and cellular telephony)

✓ Internet (fixed and mobile)

3.- Telecommunications Technologies

✓ Analog system and NTSC / DVB-S / DVB-T (Sound and television broadcasting)

✓ GSM / 2G, GPRS, EDGE, 3G, 4G (Mobile Telephony)

✓ Wimax, 4G (Internet)

4.- Telecommunications Infrastructure

✓ National Transmission Network in Fiber Optics or Microwave Links (Telephony and Internet)

✓ International access (submarine cable, satellite link for telephony and Internet)

✓ Free – to air sound and television broadcasting station, scrambled TV stations, cable and satellite TV systems

5.- Penetration of services

✓ Radio coverage of the territory or rate of deployment of optical cables

21 Politique des télécommunications https://www.ic.gc.ca/eic/site/693.nsf/fra/h_00015.html

- ✓ Percentage of the population using telecommunications services
  - o Percentage of listeners and viewers
  - o Percentage of telephone subscribers
  - o Percentage of Internet users

6.- Telecommunications Operators

- ✓ Telephony Operators, Virtual Operators, Network Operators, Data Transmission Operators, Transmission Operators
- ✓ Internet Service Providers
- ✓ Broadcasting Operators

7.- Work in progress in the world

- ✓ Transition to Digital Television
- ✓ Transition from IPV4 to IPV6
- ✓ Development of 5G

Presentation of the International Telecommunication Union (ITU)

- ✓ United Nations specialized agency for Telecommunications and Information and Communication Technologies
- ✓ Organization based on public-private partnerships
- ✓ Institution making all electronic communications possible through the management of telecommunication resources, the development of standards and the improvement of access for all
- ✓ Main international telecommunications standardization body
- ✓ Working languages: 6 languages (English, Arabic, Chinese, Spanish, French, Russian)
- ✓ Head office: Geneva, Switzerland
- ✓ Offices: 12 regional offices and area offices around the world
- ✓ Members: 193 Member States and nearly 800 private sector entities and academic institutions Sector Members

ITU History

- ✓ Founded in Paris in 1865 as the International Telegraph Union (ITU) to replace CCITT (International Telegraphy and Telephony Consultative Committee)
- ✓ Became International Telecommunication Union (ITU) since 1932
- ✓ Became United Nations Specialized Agency for Information Technology and Communication since 1947
- ✓ Fields of activity: from digital broadcasting to the Internet, via mobile technologies and TV3D[22]

ITU Administration

- ✓ Plenipotentiary Conference
  - o Supreme organ
  - o Decision-making body (general guidelines of the Union and its activities)
- ✓ ITU Council
  - o Review of major telecommunication policy issues (adaptation of activities, policy orientations, Union strategy for the dynamic telecommunications environment)
  - o Development of ITU Policy and Strategic Planning
  - o Management of the functioning of the Union
  - o Coordination of work programs
  - o Approval of budgets
  - o Control of finances and expenses[23]

ITU Organization

- ✓ General Secretariat
- ✓ Deputy Secretariat General
- ✓ Radiocommunication Bureau
- ✓ Bureau of Standardization
- ✓ Telecommunications Development Bureau

ITU Structures

- ✓ General Secretariat
- ✓ Standardization Sector (ITU -T)

22 Histoire de l'UIT
http://www.itu.int/fr/about/Pages/history.aspx

23 Présentation du conseil
http://www.itu.int/fr/council/Pages/overview.aspx

- ✓ Radiocommunication Sector (ITU-R)
- ✓ Telecommunication Development Sector (ITU -D)

ITU members

- ✓ Governments (National Telecommunications Administration)
- ✓ Major equipment manufacturers
- ✓ Global operators
- ✓ Innovative small businesses using new or emerging technologies
- ✓ Large institutes and academic research and development institutions

ITU functions

- ✓ World-wide allocation of radio frequencies and telecommunications satellite orbits
- ✓ Development of technical standards to ensure the harmonious interconnection of networks and technologies
- ✓ Definition of tariffs and accounting principles for international telecommunications services
- ✓ Improvement of access to ICTs for underserved communities

ITU activities

- ✓ Conferences
- ✓ Meetings
- ✓ Involvements in agreements related to technologies and services destined to users
- ✓ Allocation of resources for the provision of telecommunications services (radio frequencies and satellite orbital positions)
- ✓ Establishments of standards, protocols and international agreements for the provision of services
- ✓ International coordination of satellite management and spectrum and orbits (access to television, satellite navigation systems, weather reports and online maps)
- ✓ Access to telecommunications services in remote areas of the planet
- ✓ Access to the Internet (Development of the main standards of the Internet)
- ✓ Support for the provision of telecommunication services in case of disasters
- ✓ Collaboration with the private sector in the definition of new technologies for new networks and services
- ✓ Collaboration with public and private sector partners to facilitate access to ICTs and make ICTs services affordable, equitable and universal
- ✓ Provision of capacity building to world populations (ICTs use and training)[24]

ITU flagship event

- ✓ Plenipotentiary Conference (PP)
  - o Supreme organ of the Union
  - o PP convened every 4 years
  - o Participation of Member States
  - o Participation of Sector Members, regional telecommunication organizations and intergovernmental organizations, the United Nations and its specialized agencies as observers[25]

Tasks of the Plenipotentiary Conferences

- ✓ Decision on the future role of the Union
- ✓ Determination of the capacity of influence and guide of the evolution of ICTs in the world of the Union
- ✓ Determination of the general principles of the Union
- ✓ Adoption of a strategic plan and a financial plan for a period of 4 years
- ✓ Election of members of the management team of the organization, and members of the Board and members of the Radio Regulations Board[26]

Activities of the Standardization Sector (ITU - T)

- ✓ Development of international standards (recommendations) to facilitate the functioning of the

24 A propos de l'UIT http://www.itu.int/fr/about/Pages/vision.aspx

25 À propos de la Conférence de plénipotentiaires – http://www.itu.int/plenipotentiary/2010/about-fr.html

26 À propos de la Conférence de plénipotentiaires – http://www.itu.int/plenipotentiary/2010/about-fr.html

global information and communication technologies and infrastructures

- ✓ Operational aspects
- ✓ Rates and Accounting for International Telecommunication Services
- ✓ Management (use of telecommunications management network)
- ✓ Protections of telecommunications facilities from interference and lightning
- ✓ Outside network
- ✓ Integrated broadband
- ✓ Signaling and protocols
- ✓ Transport networks
  - o Optical Transport Network and Others, Systems and Equipment
- ✓ Multimedia Services
  - o Definition of multimedia, multimedia systems, terminals, protocols and signaling
- ✓ Data networks and software
- ✓ IMT - 2000[27]

ITU-T flagship event

- ✓ World Telecommunication Standardization Assembly (WTSA)
  - o WTSA convened every 4 years

Tasks of the WTSA

Review of current working methods

- ✓ Review of approval process, work program and study group structure
- ✓ Definition of general policies
- ✓ Adoption of ITU-T working methods and procedures[28]
- ✓ Definition of strategic directions for the Telecommunication Standardization Sector (ITU -T)
- ✓ Revision of the structure and working methods of ITU -T
- ✓ Review of collaboration mechanisms between the sector and other standardization bodies, SMEs and Open Source communities[29]

Fields of activity of the Radiocommunication Sector (ITU-R)

- ✓ Management of the radio spectrum at the global level
  - o Principles and techniques for spectrum management
  - o Methods and techniques
- ✓ Management of telecommunications satellite orbits
- ✓ Propagation of radio waves
  - o Propagation of radio waves in ionized and non-ionized environment
- ✓ Fixed satellite services
  - o Systems and networks for the fixed-satellite service and inter-satellite links
- ✓ Broadcasting Services
  - o Audio, video, multimedia and data services for the general public
- ✓ Scientific Services
  - o Systems for space operations, space research
- ✓ Mobile services
  - o Systems and networks for mobile and radio determination
- ✓ Fixed services
- ✓ Fixed service systems and networks operating earth stations
- ✓ Development of international standards for radiocommunication systems
- ✓ Coordination of wireless services
- ✓ Accomplissement through conferences and study groups of important works on broadband mobile communications and broadcasting techniques

27 Presentation on ITU - https://www.slideshare.net/bijen-khagi/itu-and-its-sector?next_slideshow=1

28 Assemblée mondiale de Normalisation des Télécommunications – http://www.itu.int/ITU-T/wtsa-08/index-fr.html

29 Assemblée mondiale de Normalisation des Télécommunications – http://www.itu.int/fr/mediacentre/Pages/2016-MA13.aspx

such as Ultra-wideband HDTV and TV3D

ITU-R flagship event

- ✓ World Radiocommunication Conference (WRC)
  - o WRC convened every 3 or 4 years

WRC Tasks

- ✓ Review of the Radio Regulation (International Treaty Governing the Use of the Radio Spectrum and Geostationary and Non-Geostationary Satellite Orbits)
- ✓ Revision of Assignment Plans or Associated Frequency Allotment Plans
- ✓ CReview of any world wide radiocommunication issue of global significance
- ✓ Reorientations for the Radio Regulations Board and at Radiocommunication Bureau
- ✓ Review of the activities of the the Radio Regulations Board and at Radiocommunication Bureau
- ✓ Determination of issues for consideration by the radiocommunication assembly and its study groups for future radiocommunication conferences[30]

Fields of activity of the Telecommunication Development Sector (ITU-D)

- ✓ Easy access to telecommunications
- ✓ Affordable cost for telecommunications services
- ✓ Socio-economic development
- ✓ Bridging of the digital divide[31]
- ✓ Deployment of broadband networks (strategies and policies)
- ✓ Migration of networks and interconnection
- ✓ New technologies for rural communications
- ✓ Digital broadcasting technologies[32]

ITU-D flagship event

- ✓ World Telecommunication Development Conference (WTDC)
  - o WTDC convened every 3 or 4 years

WTDC Tasks

- ✓ Review of Themes, Projects and Programs related to the Telecommunication Development
- ✓ Definition of strategies and objectives related to telecommunication development[33]

Legislative aspects of telecommunications

- ✓ Development of Telecommunications Laws
- ✓ Development of Universal Access Laws
- ✓ Law on access of persons with disabilities to telephone services

Legal framework of telecommunications

- ✓ Telecommunications/ICTs Sector Law
- ✓ Regulatory text
- ✓ Technical and legal standards
- ✓ Conditions for setting up electronic communications networks
- ✓ Conditions for the provision of electronic communications services
- ✓ Supervision of the universal service
- ✓ Warranty conditions of competition and management of radio frequency spectrum resources and numbering
- ✓ Definition of conditions of restriction for property rights,
- ✓ Definition of consumer rights
- ✓ Security conditions for networks and services and their operations in case of emergency
- ✓ Protection of the rights of consumers of public communication services to confidentiality of communications
- ✓ Dispute resolution between entities in the electronic communications market
- ✓ Management of skills, organization and operations of communication networks and services

Telecommunications Law

- ✓ Right of features, contents, principles and struc-

30 Conférences mondiales des radiocommunications (CMR) http://www.itu.int/fr/ITU-R/conferences/wrc/Pages/default.aspx

31 http://www.itu.int/fr/join/Pages/default.aspx

32 Presentation on ITU - https://www.slideshare.net/bijen-khagi/itu-and-its-sector?next_slideshow=1

33 WTDC-17 - http://www.itu.int/en/ITU-D/Conferences/WTDC/WTDC17/Pages/About.aspx

tures

- ✓ Basic elements of the legal regime of information and communications technologies
- ✓ Foundations and key aspects of the regulation of telecommunication networks and services
- ✓ Access to universal telecommunications services
- ✓ Access to other telecommunications services provided on the covered area
- ✓ Freedom of choice of telecommunications service provider
- ✓ Equal access to telecommunications services
- ✓ Access to basic information on the conditions of provision of telecommunications services and their pricing
- ✓ Obligation of all service users to comply with the regulations in force relating to the connection to public telecommunications networks[34]

Legal aspects of telecommunications

- ✓ Organization, status and competences of the regulator
- ✓ Management of Spectrum radio waves
- ✓ Infrastructure sharing
- ✓ Competition rights
- ✓ Allocation of scarce resources
- ✓ Revenue Generation for the State
- ✓ Consumer protection
- ✓ Legal security

Telecommunications Act

- ✓ Regulation of electronic communications by wire and radio
  - o Broadcasting (Sound and Television Broadcasting) Regulations
  - o Telephony
  - o Communication Services
  - o Cable television
  - o Satellite communication
  - o Wireless communication
  - o Internet

Fields of intervention of telecommunications laws

5 main fields

Regulation of the radio frequency spectrum

- ✓ Adoption of rules for spectrum management (licensing conditions)
- ✓ Rules for Assignment of Frequency Blocks for Government and Private Sector use
- ✓ Rules for the assignment of frequencies to commercial use (spectrum auction)

Market regulation

- ✓ Adoption of rules for the management of relations between the various communication industries and market players
- ✓ Adoption of rules on
  - o the obligation to carry signals
  - o retransmission, interconnection of telecommunications facilities, mobile roaming,
  - o compensation between carriers,
  - o access and programme transport by cable,
  - o network neutrality
  - o support structure of utilities

Content regulation

Adoption of laws prohibiting the dissemination of obscenity

Limitation of commercial content in children's programming

Adoption of rules to guarantee media coverage of local events

Adoption of rules to preserve diversity of viewpoints by preventing the concentration of media ownership in local markets

Access to the telecommunications market

- ✓ Establishment of rules to ensure the entry of new operators into the market

---

34 Code des télécommunications http://droitdu.net/files/sites/107/2013/11/code_des_telecommunications.pdf

- ✓ Guarantee for the deployment of telecoms infrastructure in all areas under fair conditions

Consumer protection

- ✓ Verification of the reasonableness of the rates, terms and conditions of communication services provided to the public
- ✓ Monitoring of the mandatory provision of subtitling and services to the hearing impaired
- ✓ Review of mergers and acquisitions to ensure consumer interests in consolidation[35]

Interactions between telecommunication players

- ✓ Manufacturer or supplier : design, development, production and sale of equipment necessary for the provision of telecommunications services
- ✓ Telecommunications operator: sound and rigorous use of the equipment supplied by the equipment manufacturer
- ✓ User: consumption of telecommunication services provided by the operator and supported by telecommunications equipment
- ✓ User or user: formulation of requirements and unconsciousness of technical difficulties[36]

Telecommunications and society

- ✓ Telecommunications: collective support for personal communications
- ✓ Important role of electronic communications in human exchanges
- ✓ Means of participation and involvement of all in social life
- ✓ Telecom/ICTs: Technological foundation for communications within society (functioning of families, businesses and governments based on ICTs tools)
- ✓ Telecom/ICTs: Gateway for Participation and Development
- ✓ Telecom/ICTs: Vital infrastructure for national security (assistance essential for maintaining national security, exploitation of ICTs for interventions before, during and after natural disasters, internal security of the State, communication of information related to the intelligence service and communication within the army

Contribution of telecom/ICTs to society

- ✓ Availability of remote communication services
- ✓ Platform to electronically provide services formerly traditionally provided
- ✓ Activator or lever for other services and other sectors
- ✓ Jobs creation

Social Aspects of Telecoms/ICTs

- ✓ Growth in the computer, multimedia, telephony and Internet markets
- ✓ Merger of telecommunications branches
- ✓ Emergence of new ways of communication

Problems related to ICTs development in some countries

Different obstacles to ICTs development

Obstacle 1: Legal framework unsuited to the development of the sector

- ✓ Need for a legal framework adapted to the current development of the sector
- ✓ *Legal framework*: indispensable tool for the harmonious development of the sector
- ✓ *Legal framework*: attraction of investments in the sector
- ✓ *Legal framework*: guarantee for consumers
- ✓ *Appropriate and adapted legal framework*: a tool for optimal use of telecoms / ICTs sector potentials

Obstacle 2: Lack of government ICTs policy

- ✓ Vision for the sector
- ✓ Need for a government policy for the development of the ICTs sector
- ✓ Definition of strategies to adopt

Obstacle 3: Constraints of access to ICTs services within the national territory

- ✓ Availability of all services in all areas in a non-discriminatory manner

35 Communication law https://en.wikipedia.org/wiki/Communications_law

36 Systèmes de télécommunications, base de transmission P.- G. Fontolliet

- ✓ Access and use of ICTs services
- ✓ Provision of universal access and service

Obstacle 4: Lack of awareness for ICTs use

- ✓ Lack of incentive to use ICTs services
- ✓ Adoption of new methods (abandonment of traditional methods)
- ✓ Transition to technological solutions
- ✓ Resistance to change

Obstacle 5: Morbid competitive environment

- ✓ Competition: a key element in the development of the Telecom/ICTs sector
- ✓ Competition: more services
- ✓ Competition: more technologies
- ✓ Competition: more jobs

Obstacle 6: Low purchasing power of consumers

- ✓ Low purchasing power of a high percentage of users
- ✓ Direct impact on the consumption of services
- ✓ Low consumption of services
- ✓ Incapacity of acquisition of the access terminal

Obstacle 7: High illiteracy rate

- ✓ Basic level of knowledge for use of ICTs services and applications
- ✓ Reading and writing ability for basic services
- ✓ Initiation to ICTs for advanced services (Internet, social networks, etc.)
- ✓ High illiteracy rate = brake on ICTs development

International aspects of telecommunications

- ✓ Broadcasting of television and radio programs of a country in other neighboring countries
- ✓ Installation of submarine cables between two countries (in maritime waters)
- ✓ Management of radio interference in border areas (borders crossing by waves)
- ✓ International frequency coordination (fight against harmful radio interference)
- ✓ Radiation of satellite signals on neighboring territories
- ✓ Interconnection of networks from different countries for international traffic routing
- ✓ International Roaming
- ✓ International Traffic settlement between 2 countries (International Trade)
- ✓ Involvement of international organizations and institutions (WTO, ITU, ICANN, IGF) in telecommunication management

International Telecommunications

- ✓ *International telecommunications service*: provision of telecommunications between offices or telecommunication stations of all kinds, located in different countries or belonging to different countries
- ✓ *International route*: means and technical facilities, located in different countries, used for the transport of telecommunication traffic between two international telecommunication terminal centers or offices
- ✓ *Accounting rates*: price set by agreement between authorized operations, for a given relationship and used for the establishment of international accounts
- ✓ *Collection* costs: fees established and collected by an authorized operation from its customers for the use of an international telecommunications service[37]

Actors involved in international telecommunications

- ✓ International Telecommunications Union
    - o Telecoms Resource Management
    - o Standards development
    - o Negotiations between countries
- ✓ World Trade Organization
    - o Extension of the GATT agreement to telecommunications

37 Règlement des télécommunications internationales - Extrait de la publication : Actes finals de la Conférence mondiale des télécommunications internationales (Dubaï, 2012) (Genève: UIT, 2013) - http://search.itu.int/history/HistoryDigitalCollectionDocLibrary/1.42.48.fr.201.pdf

- o Conflict resolution
- ✓ Standardization bodies
  - o Standards for the manufacturing of equipment (terminals and systems)
  - o Compatibility of equipment used worldwide

REGULATION OF THE TELECOMMUNICATIONS SECTOR

- ✓ Set of legal, economic and technical measures to promote the exercise of telecommunications activities
- ✓ Set of legislative and regulatory texts applicable to the telecommunication sector[38]
- ✓ Actions and decisions to ensure the dynamic evolution of the sector[39]
- ✓ Application of the regulatory framework defined by the public authorities[40]

Objectives of the regulation

- ✓ Protection of the interests of the State
- ✓ Protection of the interests of the operators
- ✓ Protection of the interests of the consumer

Need for regulation of the telecommunications sector

- ✓ Avoiding the failure of the telecommunications market
- ✓ Encouraging effective competition
- ✓ Protecting the interests of consumers
- ✓ Increasing access to technology and services[41]

Foundations of regulation

- ✓ Competition Policy for Access to Telecommunications Services
- ✓ Scarcity of radio frequency spectrum resources
- ✓ Development of standards for consumer protection
- ✓ Protection of private life[42]

Regulation

- ✓ Application tools for the management of the sector
- ✓ Legal instrument, with legal force, adopted by an authority under a law and prescribing standards of conduct[43]
- ✓ Tool for the application of telecommunications policy

Telecommunications Regulatory Authority (Regulator)

- ✓ State organ responsible for monitoring and managing the telecommunications sector
- ✓ Authority with regulatory, arbitration, control and sanction powers in the telecommunications sector

Missions of the Regulator

2 main missions

1.- Activities of regulation of the telecommunications market

2.- Activities of development of the telecommunications sector

Actions arising from these missions

- ✓ Adoption of specific regulations
- ✓ Monitoring of applications by operators the specifications and other rules established in accordance with the law
- ✓ Compliance with technical regulations in the telecommunications sector including broadcasting
- ✓ Protection of the interest of consumers and the general population
- ✓ Arbitration of disputes between operators according to the procedures defined by law

Legal measures

Management of certain aspects of the mission based

38 Régulation des télécommunications, article de référence, Alain Vallée

39 Régulation des télécommunications, article de référence, Alain Vallée

40 Notes d'Oumar Kane sur la régulation et télécommunications

41 Telecommunication manual handbook, tenth anniversary edition - https://www.infodev.org/infodev-files/resource/InfodevDocuments_1057.pdf

42 Media, Communications and the Internet - The Regulatory Framework by John Corker http://slideplayer.com/slide/5372381/

43 Notes d'Oumar Kane sur la régulation et télécommunications

on ex ante and ex post regulation

Technical regulation

- ✓ Definition of technical standards
- ✓ Definition of standards and control of compliance with these standards
- ✓ Homologation and control of telecommunications equipment
- ✓ Licensing for the use of radiocommunication equipment
- ✓ Examination of applications for authorization for the provision of telecommunications services (telephony, Internet, broadcasting, data transmission, radiotelephony, etc.)
- ✓ Management and control of the radio frequency spectrum
- ✓ Management of the national numbering plan
- ✓ Management of high points (main radiocommunication sites) and control of earth stations
- ✓ Quality control of services provided by operators

Economic regulation

- ✓ Guarantee of fair competition
- ✓ Validation of operators' interconnection catalogs and tariffs
- ✓ Monitoring of the prices of services offered by operators
- ✓ Monitoring of tariffs practiced in telecommunications networks
- ✓ Reinforcement of compliance for fair and healthy competition
- ✓ Economic monitoring of operators' obligations: collection of fees for authorizations granted for the marketing of telecommunications services
- ✓ Promotion of universal access and reasonable cost practice
- ✓ Supervision of operators' activities and adoption of regulatory decisions (interconnection, pricing, traffic, etc.)
- ✓ Promotion of job creation and sustained growth of the economy through Information and Communication Technologies (ICTs)[44]

Players under regulation in the Telecommunications field

Players subject to regulation

- ✓ Operators of broadcasting (sound and television)
- ✓ Carriers (transport operators)
- ✓ Fixed and mobile Operators
- ✓ Internet Service Providers
- ✓ Telecommunications Satellite Operators
- ✓ Satellite Operators

Objects of regulation in telecommunications

- ✓ Signal transport
  - o Definition of the conditions of transport / broadcasting / transmission
  - o Access to telecommunications and service infrastructures
- ✓ Consumer protection
  - o Basic standards for equipment
  - o Behavior of the service provider
- ✓ Protection of the interests of the State
  - o Management of State resources used
  - o State revenues protection

Telecommunications Regulatory Methods

- ✓ Direct regulation (laws, regulations, standards, licensing with conditions attached)
- ✓ Co-regulation (code of good practice approved or endorsed by the government or the regulator)
- ✓ Self-regulation (code of good practices endorsed by the industry)
- ✓ Regulation dictated by economic and technological methods[45]

Telecommunications Regulatory Generations

- ✓ G1: Regulated public monopolies
  - o Independent regulator

44 Régulation du secteur des Télécommunications en Hatti – Conseil National des Télécommunications

45 Methods of regulation

- o Traditional regulatory approach (coercive approach)
- ✓ G2: Basic Reforms (Opening of Telecommunications Markets)
  - o Creation of a separate regulatory authority
  - o Partial liberalization
  - o Privatization
- ✓ G3: Regulation focused on investment, innovation and access
  - o Double focus on competition stimulation in service and content delivery and consumer protection
- ✓ G4: Integrated regulation
  - o With an evolving role of the regulator as a partner for development and social inclusion
  - o Regulation driven by economic and social policy objectives
- ✓ G5: Collaborative regulation
  - o Need to define foundation, platforms and mechanisms for cooperation with regulators in other sectors in order to reach the sustainable development goals
  - o Inclusive Dialogue and Harmonized Approach between Sectors[46]

Forms of Regulatory Authority

- ✓ Regulation exercised by a single regulator (under the supervision of a ministry, or independent)
- ✓ Regulation exercised by several entities
  - o A regulatory authority
  - o A national frequency management agency
  - o An Audiovisual Council (Radio and Television

Regulatory activities

- ✓ Protecting consumer interests
- ✓ Ensuring respect for the privacy of consumers
- ✓ Ensuring an honest and effective competition
- ✓ Ensuring a good level of services at affordable prices throughout the national territory
- ✓ Encouraging the public use of telecoms services as a support infrastructure for all levels of economic and social development of the population
- ✓ Ensuring the efficient and interference-free use of radio spectrum for telecoms services including radio and television and all other services made available by ICTs.
- ✓ Ensuring the availability of services under a free competition regime

Challenges of regulation

- ✓ Constant changes in technologies
- ✓ Content digitization and digital transmission
- ✓ Growth in the use of the Internet
- ✓ Liberalization of telecommunications markets
- ✓ Monopolies on infrastructures
- ✓ Change in delivery methods
- ✓ Growth in the media niches
- ✓ Interactivity of the media
- ✓ Ease of publication and distribution of individuals[47]

STANDARDS IN TELECOMMUNICATIONS

- ✓ Standard : Tool favoring the operation of a system composed of different elements (transmission media, switches, modems, hubs, routers, terminals, software and programming languages, operating systems ... ..)
- ✓ Standard: set of technical rules to promote interoperability of different systems
- ✓ Standard: technical requirements and specifications for the manufacturing and operation of

46 Chapter 5: Spanning the Internet divide to drive development (ITU) https://www.wto.org/english/tratop_e/devel_e/a4t_e/4sers_2_vanessa_gray_itu_chapter_5_aft_2017_aid_for_trade_presentation_may30.pdf

47 Media, Communications and the Internet - The Regulatory Framework by John Corker http://slideplayer.com/slide/5372381/

equipment, or a system as a whole[48]

- ✓ Standard: an essential tool for accessing and using telecommunications services (telephone calls, SMS, e-mail, Internet browsing, etc.)
- ✓ Standard: guarantee of interoperability between telecommunications systems and equipment
- ✓ Standard : Operating and interoperable terms of telecommunication networks[49]

Roles of standards in the telecommunications and ICTs sector

- ✓ Allowing exchanges between machines of different manufacturers
- ✓ Ensuring the independence of applications with respect to transmission constraints
- ✓ Ensuring future evolution without compromising the software and hardware architecture[50]

Standardization of the Telecommunications Sector

- ✓ Strategy for compatibility and interworking of systems and equipment manufactured by different manufacturers
- ✓ Means of promoting compatibility between different telecommunication systems in different regions of the world
- ✓ Determination of the interfaces to be installed between the different devices to facilitate communication between them
- ✓ Description of the functions of each equipment

Types of standards in telecommunications

- ✓ Equipment manufacturing standards
- ✓ Operating Standards for Telecommunications Systems
- ✓ Taxation and billing standards
- ✓ Frequency allocation standards
- ✓ Electromagnetic radiation standards[51]

International Telecommunication Standardization Organizations

- ✓ ITU: International Telecommunication Union
- ✓ ISO: International Standardization Organization
- ✓ IEC: International Electrotechnical Commission
- ✓ ETSI: European Telecommunication Standards Institute (regional body, Europe)
- ✓ ANSI: American National Standard Institute (National Organization, United States of America)
- ✓ IEEE: Institute of Electrical and Electronics Engineers (National Organization, United States of America)
- ✓ IETF: Internet Engineering Task Force

National Standardization Organizations

- ✓ AFNOR: French Association of Normalization
- ✓ ANSI: American National Standard Institute
- ✓ BSI: British Standard Institute
- ✓ DIN: Deutsches Institut Für Normung

ITU standards

- ✓ **ITU standards(called"Recommendations")** : Fundamental tools for the operation of telecommunication networks and ICTs systems
- ✓ More than 4000 ITU standards (recommendations) in force around the world for:
  - o Definition of services
  - o Architecture and network security
  - o Broadband digital subscriber lines
  - o Optical transmission systems (Gbit / s)
  - o Next Generation Networks (NGN)
  - o IP issues[52]

Players involved in standards development

Equipment suppliers

- ✓ Regulators
- ✓ Network Operators
- ✓ Service Providers

48 Université des Frères Mentouri Constantine 1 http://www.umc.edu.dz/images/UEF2.2.1.pdf

49 Recommandations et autres publications de l'UIT-T https://www.itu.int/fr/ITU-T/publications/Pages/default.aspx

50 Réseaux - Formation Télécom Réseaux Pléneuf

51 Telecommunications Law - Ian Lloyd and David Mellor

52 Recommandations de l'UIT-T http://www.itu.int/fr/ITU-T/publications/Pages/recs.aspx

- ✓ Standardizations bodies

Examples of standards

- ✓ ITU -T E.164 : Recommendation defining the structure of a telephone number and fixing at a maximum of 15 digits its length
- ✓ ITU - T G. 711 : European standard for traditional fixed telephony
    - o Speed : 64kb/s
- ✓ ITU-T G.729: standard (recommendation) applied to voice over IP
    - o Speed : 8 kbit/s
- ✓ Recommendation ITU T H.264: widely used standard for video compression
- ✓ **IEEE 802.11**: Wi-Fi Access Standard (a standard developed by the Institute of Electrical and Electronics Engineers (IEEE) for access to wireless Internet via a local computer network)
- ✓ H.261: standard used for videoconferences (between 40 Kbps and 2 Mbps)
- ✓ 802.3 for Ethernet networks,
- ✓ 802.4 for Token Bus networks
- ✓ 802.5 for Token Ring

OVERVIEW OF THE TECHNICAL EDUCATION OF THE SECTOR'S HUMAN RESOURCES

- ✓ *Basic subjects*
    - o Mathematics
    - o Physical
    - o Chemistry
    - o Computing
- ✓ *Specialization subjects*
    - o Electromagnetism
    - o Electrotechnics
    - o Analog and digital electronics
    - o Component technologies
    - o Signal Analysis and Systems
    - o Theory of information
    - o Telecommunications Systems and Network Interoperability
    - o Analogue and digital communication systems
    - o Broadcasting
    - o Transmission system
    - o Operating systems
    - o Web and Internet
    - o Web programming
    - o Antennas and propagation
    - o TCP/IP protocol
    - o Network standards and interconnections
    - o Analog and digital modulations
    - o Programming
    - o Mobile telephony
    - o IP telephony
    - o Data transmission

- o Satellite transmission
- o Network security
- o Optical communication system
- o High Frequencies Transmission Lines
- o Digital transmissions
- o Servo Systems
- o Analogue and digital television
- o Sensors

TELECOMMUNICATIONS SECTOR JOBS

- ✓ Deputy Project Manager - Telecoms New Services
- ✓ Network Administrator - Telecom
- ✓ Call Center Technical Administrator
- ✓ Administrator of Virtualization of Information Systems
- ✓ Technical Architect -Telecoms
- ✓ Manager
- ✓ Project Manager of Cloud Computing
- ✓ International Project Manager in Computer Science and Networks
- ✓ Project Manager New Services -telecom
- ✓ Telecoms Project Manager
- ✓ Commercial Telephony - B to B -
- ✓ Designer of Mobile Applications
- ✓ Telecommunications Consultant
- ✓ Interactive Designer
- ✓ IOS developer
- ✓ Mobile Developer
- ✓ Call Center Director
- ✓ Electronics - Cable Operator Assembler
- ✓ Expert in Telecoms and Networks
- ✓ Facilitator
- ✓ Hotliner
- ✓ Telecoms Engineer - Specialization Satellite Communications
- ✓ Telecommunications Business Engineer
- ✓ Engineering and Development Engineer - Computer Science
- ✓ Engineer of the Internet
- ✓ Telecom Research Engineer
- ✓ Electronics Engineer
- ✓ Geomatics Engineer
- ✓ Metrology Engineer - Embedded Systems
- ✓ Telecommunication engineer
- ✓ Operations Engineer Telephony Network
- ✓ Computer engineer in charge of tests
- ✓ Network Engineer -Telecom
- ✓ Network Engineer - Option Mobiles
- ✓ Cisco Systems Network Engineer
- ✓ Customer Care Engineer
- ✓ Corporate accounts Engineer
- ✓ Telecoms Engineer in charge of testing
- ✓ Voice Engineer
- ✓ Research Engineer
- ✓ Cable Communications System Installer
- ✓ Engineer Manager
- ✓ Network Studies Manager (IT)
- ✓ Telecommunications Manager
- ✓ Computer Maintenance Technician
- ✓ Technician of the wired communication networks
- ✓ Electronics Technician
- ✓ Intrusion Monitoring Facility Technician
- ✓ Technician in optronics
- ✓ Network Operator Technician -Telephony

- ✓ Network and Telephony Technician
- ✓ Network-messaging technician
- ✓ Network Technician - Mobile Telephony
- ✓ Network and Corporate Telecommunications Technician
- ✓ Senior Technician in Computer Networks and Telecommunications
- ✓ Senior Technician Specialized Networks-Telecoms Services
- ✓ Senior Technician - Telephony Network Operations
- ✓ Mobile Telephony consultant[53]

53 Secteur des télécommunications - Secteur d'activité : Télécommunications
http://www.leguidedesmetiers.com/formations-et-metiers/secteur-telecommunications/12

CHAPTER 3

TELECOMMUNICATIONS SERVICES AND THEIR USES

TELECOMMUNICATIONS SERVICES (ELECTRONIC COMMUNICATIONS SERVICES)

Telecommunications Service

- ✓ Transport of the user's message (information) from point A to point B using electronic communications systems
- ✓ Intangible product consisting of a temporal use of a terminal attached to an electronic communications infrastructure
- ✓ Paid service consisting primarily of sending signals over electronic communications networks, including telecoms and data transmission services in networks used for broadcasting purposes[54]
- ✓ Service of signals transmission through electronic communications networks
- ✓ Service provided through telecommunications facility

Framework for the Provision of Telecommunications Services

- ✓ Technical principles
  - o Application of physical laws (electricity, electronics)
  - o Design, development, deployment and operation of systems
- ✓ Laws (legal framework)
  - o Conditions of licensing
  - o Resources usage conditions
  - o Investments protection
  - o Consumer protection
  - o Environment conducive to fair competition
- ✓ Economic model
  - o Purchase of telecommunications equipment
  - o Sales of services and products
  - o Royalties for used resources

54 Réseaux et Télécommunications - Dominique SERET, Ahmed MEHAOUA, Neilze DORTA http://www.mi.parisdescartes.fr/~mea/cours/L3/L3.poly06.pdf

Bases of the framework for Telecommunications Service Delivery

- ✓ Technical principles
- ✓ Standards
- ✓ Policies
- ✓ Constraints[55]

Steps in the provision of Telecoms / ICTs services Design

- ✓ Development
- ✓ Deployment
- ✓ Operation
- ✓ Suspension[56]

Bases of the availability of telecommunications services

- ✓ Preventive maintenance of telecommunications and computer equipment
- ✓ Availability of human resources to manage operations
- ✓ Supply of uninterrupted electrical power

Telecommunications Services and Users

- ✓ Interactive services
  - o Telephony, videoconferencing, video telephony, electronic mail, consultation of documents, images and videos
- ✓ Broadcasting Services
  - o Radio, television[57]

Types of telecommunications services

- ✓ One way Audiovisual Services: Sound and Television Broadcasting
- ✓ Bilateral services: Telephony, Videoconferencing, Instant Messaging
- ✓ Data services: data transmission, Internet

55 Service Delivery Model https://wiki.doit.wisc.edu/confluence/display/MADLIB/Service+-Delivery+Model

56 Service Delivery Model https://wiki.doit.wisc.edu/confluence/display/MADLIB/Service+-Delivery+Model

57 Principes de Base http://www.httr.ups-tlse.fr/pedagogie/annexes/intro/principes.pdf

Categories of Telecommunications Services

2 categories of telecommunications services : Basic Telecommunications and value - addedservices

1.- Basic telecommunications

- ✓ Transport of voice signals or data from a point of departure to the point of arrival
- ✓ Public or private communication services
- ✓ End-to-end transmission of information provided by the customer

Examples of basic telecommunication services

- ✓ Telephone Services
- ✓ Packet switched data transmission services
- ✓ Circuit switched data transmission services
- ✓ Telex services
- ✓ Telegraph services
- ✓ Fax Services
- ✓ Private leased circuit services
- ✓ Other (Cellular/Mobile Analogue/Digital Services
  - o Mobile data services
  - o Radio Research Services
  - o Personal Communications Services
  - o Mobile satellite services (including, for example, telephone, data, paging and / or personal communications services)
  - o Fixed satellite services
  - o VSAT Services
  - o Access earth station services
  - o Teleconferencing Services
  - o Video transmission services
  - o Shared Resource Radio Services

2.- Value-added services

- ✓ Value added by service providers to customer-provided information
- ✓ Improvement of the forms of information provided by the client
- ✓ Improvement of the contents of the information provided by the client
- ✓ Means of storing information
- ✓ Means of research on information

Examples of value-added services

- ✓ Online data processing services
- ✓ Online storage and search services in databases
- ✓ Electronic Data Interchange Services
- ✓ Email Services
- ✓ Voice Mail Services[58]
- ✓ Other value-added services
  - o Cell phone account balance
  - o Possibility of recharge 24/24
  - o Caller ID
  - o SMS-based services
  - o Sale of insurance service

Types of telecommunications services

Distinction of telecommunication services in different ways

- ✓ Type of information transmitted
- ✓ Number of partners involved
- ✓ Role played by the partners (mode of communication)

- o Unilateral: From source to recipient (Monologue)
- o Bilateral: Two-way communication (dialogue)
- o Multilateral: Exchange between several sources and several recipients (conference)[59]

Electronic communication means available to the consumer

The most known services

- ✓ Fixed and mobile telephony
- ✓ Email

58 Définition des télécommunications de base et des services à valeur ajoutée https://www.wto.org/french/tratop_f/serv_f/telecom_f/telecom_coverage_f.htm

59 Systèmes de télécommunications, Bases de transmission P.- G. Fontolliet

- ✓ Fax
- ✓ Telegraphy
- ✓ Telex
- ✓ Teletex
- ✓ Sound broadcasting
- ✓ TV broadcasting
- ✓ Website
- ✓ Personal blogs
- ✓ Instant messaging
- ✓ SMS/MMS
- ✓ Comment space in online newspapers
- ✓ Social networks
- ✓ Walkie Talkie and Citizen's Band (BC)
- ✓ Virtual Forums
- ✓ Video conference
- ✓ Online comment spaces

Classification of telecommunication services according to the nature of the messages

Sound

- ✓ Telephone
- ✓ Intercom
- ✓ Voice Messaging
- ✓ Search for people
- ✓ Conference call
- ✓ Telephone information (Speaking clock, weather)
- ✓ Broadcasting
- ✓ Mobile telephony

Texts

- ✓ Telex, teletex
- ✓ E-mail
- ✓ Electronic documentation
- ✓ Videotex
- ✓ Fax

Images

- ✓ Still image transfer
- ✓ Television
- ✓ Video Calling
- ✓ Videoconference
- ✓ Videocommunication over a wired network

Teleinformatics

- ✓ Telemetry
- ✓ Data transport
- ✓ Remote monitoring
- ✓ Remote control
- ✓ Paging[60]

Terminals for users of telecommunications systems

Electronic devices for accessing and using electronic communications services

- ✓ *Telephone*: wired or wireless communication device for transmitting the human voice remotely
- ✓ *Radio receiver (radio set)*: Electronic device designed to collect radio waves emitted by radio transmitters
- ✓ *Television receiver*: Electronic device capable of displaying images on screen the television signals received by cables or electromagnetic waves
- ✓ *Modem*: Device enabling a computer (any digital terminal) to transmit information via a telecommunications system (Interface providing conversion of digital signals in the form of analog waves, and vice versa, to ensure the exchange of information between a computer and a telecommunications system)
- ✓ *Computer*: Electronic machine designed for automated data processing.
- ✓ *Digital tablet (touch pad, tablet, digital slate)*: a slim, keyboard-less laptop with a touch screen providing nearly the same functionality as a personal computer

60 Introduction aux télécommunications http://www.volle.com/ENSPTT/introtcom.htm

- ✓ *PDA (Personal Digital Assistant)*: Handheld originally designed for organizational purposes. Services provided by a PDA: calendar, task manager, address book and e-mail software, Internet access and music or image files
- ✓ *Mobile radio receiver (mobile radio transceiver)*: a radiotelephone device capable of communicating alternately (in semi-duplex mode) while moving on foot (Example: Walkie-talkie or walkie-talkie, push to talk)
- ✓ *Set - top box*: set-top box, adapter converting a received signal into a content and displaying it on a TV screen
- ✓ *Fax*: Device designed to send and receive written documents
- ✓ *Telegraph*: Device for transmitting messages called telegrams using coded signals (usually Morse signals)

Interfaces between users and telecommunications systems

Interfaces : technical solutions

- ✓ Human communication (user): sound, speech, image, video, text and data
- ✓ Telecommunications systems : exclusive use of electrical signals, electromagnetic waves and optical signals
- ✓ Challenge met by interfaces between man and the telecommunications system
- ✓ Interface: Device interposed between the user and the telecommunications system
- ✓ Interface: Access and use of telecommunications services
- ✓ Interface: 2 missions
  - o First mission: Change of nature of consumer information for use by the telecommunications system (Transmit side)
  - o Second mission: Restitution of the consumer information in the original form (Receive side)
- ✓ Interface: conversion (on transmit side) of consumer information in a form usable by the telecommunications system
- ✓ Interface: conversion (on receive side) of the signals from the telecommunications system in a form usable by the consumer
- ✓ Interfacing between users and telecommunications systems based on energy or signals transduction.

Transduction

- ✓ Principle of transformation of one energy into another energy
- ✓ signal conversion provided by transducers

Transducer

- ✓ Device converting a physical quantity into another physical quantity
- ✓ Integrated element in the consumer terminal (in the interface between the human and the telecommunications system) for two-way conversion purposes

Transducers for sound and human speech

- o Microphone: transducer converting acoustic waves into electrical signal
- o Speaker: Transducer converting an electrical signal into acoustic waves

Transducers for image and video

- o Camera and video camera: transducer converting still and moving images into electrical signals
- o Screen: Transducer converting electrical signals into image and video usable by human being

Transducers for text and data

- o Keyboard: transducer converting text and data into electrical signal
- o Screen: Transducer converting the electrical signal into text and data usable by human being

Availability of telecommunications services

- ✓ Telephone in standby mode (network beacon channel maintaining the connection with the telephone terminal)
- ✓ Starting of the radio or television set for the reception of signals emitted by radio and television stations (radio and television stations transmitting continuously)

- ✓ Starting of the computer for access to various services (sounds, voices, images, videos, data)

Applications of electronic communications

Simplex applications

- ✓ AM and FM broadcasting
- ✓ Digital radio
- ✓ TV broadcasting
- ✓ Digital TV
- ✓ Cable TV
- ✓ Fax
- ✓ Wireless remote controls
- ✓ Paging Services
- ✓ Navigation and Direction Finding Services
- ✓ Telemetry
- ✓ Remote monitoring
- ✓ Musical services
- ✓ Internet radio and television

Duplex applications

- ✓ Telephony
- ✓ Video conference
- ✓ Instant messaging
- ✓ Two -way radio
- ✓ Radar
- ✓ Sonar
- ✓ Amateur Radio
- ✓ Citizen's bands
- ✓ Internet
- ✓ Local, wide, metropolitan networks

Needs of telecommunications users

Different telecommunications needs at home, in the office and on the move

Television

- ✓ Choice of a bouquet of channels according to the interests of the user
- ✓ Viewing of movies on demand (VOD)
- ✓ Recording of programs

Fixed telephone

- ✓ Phone call
- ✓ Voice Messaging
- ✓ Call forwarding

Mobile telephony

- ✓ Telephone call
- ✓ SMS and MMS exchange
- ✓ E-mail exchange
- ✓ Sending of pictures
- ✓ Downloading of video or music
- ✓ Access to TV
- ✓ Videoconference
- ✓ Internet browsing

Internet

- ✓ E-mail exchange
- ✓ Internet browsing
- ✓ Online shopping
- ✓ Viewing of TV and videos
- ✓ Downloading of music or movies
- ✓ Online Games
- ✓ Videoconference[61]

Triple Play

Access to *three services via a single connection (one operator)*

- ✓ Telephony
- ✓ Television
- ✓ Internet

Quadruple Play

Access to four services via a single connection (one operator)

- ✓ Broadcasting of TV programs on any terminal (GSM, TV screen, computer screen, etc.)

61 Guide pratique http://www.mediateur-telecom.fr/ressources/media/files/Guide_pratique_chapitre02.pdf

- ✓ Access to these services anywhere in the world
- ✓ Videotelephony on a specific terminal or computer, TV screen, GSM, etc.
- ✓ Internet access with a computer, TV screen, GSM, etc.[62]

TELECOMMUNICATION SERVICES FOR PERSONS WITH DISABILITIES

Services designed and implemented to facilitate electronic communications:

- o From people with disabilities to people with disabilities
- o From people with disabilities to people without disabilities

Framework of telecommunications services for persons with disabilities

- ✓ Rights of people with disabilities
- ✓ Accessibility for all to services
- ✓ Development of technical standards
  - o For the manufacturing of terminals for the persons with disabilities
  - o For the design and delivery of telecommunications services for persons with disabilities

Telecommunication Services for Persons with Disabilities (Deaf and Hard of Hearing)

Main options

- ✓ Telescription device for the hearing impaired (TTY: teletypewriter)
  - o Exchange of texts between two disabled people via the telephone network
- ✓ Relay Service (Telecommunications Relay Service)
  - o Electronic communication between a disability and a person without a disability using a relay
  - o Principle: Translation of repeated words, texts entered and videos (signs) by an operator for the two correspondents
- ✓ Subtitling (Television)
  - o Transcription of sounds into text on TV screen

Main options of relay services

- o Text relay service
- o Text relay service with VCO (Voice Carry Over)
- o Telephone Relay Service with subtitling
- o Video Relay Service[63]

Use of telecommunications

- ✓ Real time communications (telephone calls, videoconferences, etc.)
- ✓ Delayed communications (SMS, email, etc.)
- ✓ Online entertainment (movies, online games, etc.)
- ✓ Access to other services (emergency, research, online banking, etc.)

Activities dependent on telecommunications

Total dependence of operations of certain activities on telecommunications systems

- ✓ Army
- ✓ Police
- ✓ Emergency services
- ✓ Banks operations and Commerce
- ✓ Aviation and Aeronautics
- ✓ Maritime navigation

Telecommunications users and services

2 basic characteristics of users

- ✓ Unconsciousness of technical difficulties related to the provision of services
- ✓ Requirements for services provided

Man and telecommunications systems

2 interventions of the human being : Use of informations

- o Source of infirmation
- o Recipient of the information transmitted
- ✓ Human being: Control element of the telecoms system by its information (signals)
  - o Dictatetion of orders to the system to obtain ap-

62 Les enjeux de la télévision numérique http://www.awt.be/web/img/index.aspx?page=img,fr,tel,020,030

63 Des services relais pour les personnes malentendantes https://itunews.itu.int/fr/NotePrint.aspx?Note=1468

propriate answers

o Use of a telephone network[64]

Conditions of usage of telecommunications services

3 fundamental elements for the use of services: Access, Use and Skills

Access to telecommunications services

- ✓ User terminal (radio or TV, telephone, computer)
- ✓ Connection (wired or wireless connection between the terminal and the network)
- ✓ Network (system providing and managing the service)

Indicators of access

- ✓ Number of fixed telephone subscriptions per 100 inhabitants
- ✓ Number of mobile cellular subscriptions per 100 inhabitants
- ✓ International Internet bandwidth (bit/s) per user
- ✓ Percentage of households with a computer
- ✓ Percentage of households with Internet access

Conditions of use of telecommunications services

- ✓ Acquisition of the terminal
- ✓ Network Connections
- ✓ Pay-as-you-go systems or monthly subscription

Indicators of use

- ✓ Number of fixed telephone subscriptions per 100 inhabitants
- ✓ Number of mobile cellular subscriptions per 100 inhabitants.
- ✓ Number of subscriptions to the wireless broadband network per 100 inhabitants

Competence in the field of telecommunications

- ✓ Level of academic training
- ✓ Introduction to Computer Science and the Internet

64 Systèmes de télécommunications, Bases de transmission P.- G. Fontolliet

Indicators of competence

- ✓ Adult literacy rate
- ✓ Gross enrollment ratio in secondary education
- ✓ Gross enrollment rate in higher education[65]

Types of access to telecommunications services

- ✓ Wired access: connection of the user to the network by a cable
  - o Metallic Cables: Twisted Pairs, Coaxial Cables
  - o Optical Fiber
- ✓ Wireless access: End-user wireless connections to the core network
  - o Broadcasting by radio waves (AM / FM Radio, Television)
  - o Microwave link (point-to-point, point-to-multi-point links)
  - o Satellite link (radio, television, telephony, etc.)

Types of wireless access

- ✓ Fixed wireless access
- ✓ Mobile wireless access

Mechanisms for wireless service provision

- ✓ Terrestrial links
- ✓ Satellite links

Radiocommunication Services

Services provided by means of radio links

- ✓ Fixed wireless service
- ✓ Mobile wireless service
- ✓ Mobile Satellite Service
- ✓ Fixed satellite service
- ✓ Etc.

Connection and use of telecommunications services

- ✓ Connection and use of services based on an exchange of information (frequences, telephone numbers, username and password) with the system

65 Rapport: Mesurer la société de l'information 2014 https://www.itu.int/dms_pub/itu-d/opb/ind/D-IND-ICTSSSOI-2014-SUM-PDF-F.pdf

For Human Users: Interpretation and use of names

✓ Examples

o Super Star Radio

o Super action TV

o Website of the Association of Gifted Men

For Machines, servers and systems: interpretation and use of numbers (phone numbers, frequencies, Internet Protocol)

✓ Examples

o 90.5 MHz: Frequency of the Super Star radio

o Channel 8: Frequency band corresponding to the TV super Action

o + 5093415 0000: Institution X telephone number

o 192.167.113.9: "IP of the **xyz** website"

Access and use of telecommunications services

✓ Reception: adjustment of the receiver to the frequency of the transmitter (radio. TV.)

✓ Telephones already tuned to the frequency bands of telephone operators

Access and Uses of Telecommunications Services

Physical and logical connections

Logical connection

✓ Selection of the user (radio or television station) or entry of the user's personal information (user name and password for connection to computer networks, e-mail, social networks, etc.)

Physical connection

✓ (Signal availability): Receiver connected (in an area covered by radio and television station, telephone network, Internet)

✓ Connection of the user's terminal to the network (television, telephony, Internet)

Access and use based on connection

✓ *Radio*: adjustment of the receiver to the frequency of the desired station (connection type)

✓ *Television*: adjustment of the receiver to the channel (frequency band) of the selected channel (connection Type)

✓ *Fixed telephony (wired and wireless)*: Unhook of the handset switch to request the network connection

✓ *Cellular telephony*: Dialing and sending of the telephone number to request a connection

✓ *Computer networks*: Prior login (username and password) before accessing services

✓ *Email*: Entry of Username and password

✓ *Social networks*: Entry of username and password

Disconnection of telecommunications services

✓ Shutdown of the radio or television receiver

✓ Handset on-hook after a telephone conversation

✓ Disconnection due to distance from the system (weakened signal)

✓ Voluntary disconnection of a network (email, social networks, forums, computer networks, etc.)

Perceptions of the consumer of telecommunications services

✓ Service (access and use of the telecommunications service)

✓ Permanent availability

✓ Complex terminals and systems

Consumer expectations

✓ Reliability

o Quality of equipment used and permanence of services

✓ User-friendliness of the terminals

o simple to use, robust, subject to unwanted manipulation, minimum learning prior to their use

✓ Respect of the secrecy of the communications guaranteed

Specific user requirements

✓ *Mobility*: access to services everywhere

✓ *Reliability*: availability of services any time

✓ *Multimedia*: ability to access all types of information (voice, images, videos and data) via the same terminal

✓ *Broadband*: *High* speed for transferring large volumes of information

- ✓ *Portability*: Provision of portable terminals for access and use of services everywhere
- ✓ *Value for money*: quality services required for every cent invoiced

Consequences of requirements for telecoms operators

- ✓ Upgrade of the network: Increase of system capacity
  - o Use of new technologies
  - o Network expansion
- ✓ Infrastructure improvements
  - o Continued deployment of new infrastructure for services reliability

Key pillars of customer experience in the telecommunications sector[66]

4 main experiences

1.- Experience of the telecommunications network

- ✓ Coverage (signal range, signal availability in certain areas, areas covered)
- ✓ Signal quality (signal strength versus receiver threshold)
- ✓ Bit rate (signal access speed)
- ✓ Reliability (permanent availability)

2.- Commercial experience

- ✓ Price of services (affordability)
- ✓ Offers (different options, tailored solutions, etc.)
- ✓ Marketing (communication of the offer to the client, precision and clarity of the message
- ✓ Invoicing (transparency and accuracy, reflection of real consumption)
- ✓ Payment (Methods and means of payment of invoices)

3.- Experience of telecommunications product

User-friendliness of the terminal allowing access to the service

- ✓ Telephone
- ✓ Fax machine
- ✓ Modem
- ✓ TV

4.- Experience of the service

- ✓ Relationship between customer service and the customer
- ✓ Processing of the consumer requests
- ✓ Provision of appropriate answers
- ✓ Operation of the free-service
- ✓ User-friendliness of interfaces of self-service
- ✓ Accuracy of answers provided by self-service

Factors influencing the customer experience

1. Coverage, accessibility and network quality
2. Customer service
3. Multiplicity of technologies
4. Quality of terminals
5. Tripartite relationship: Operator - Regulator - Subscriber (Consumer)

Universal access

- ✓ Possibility for everyone to access the service somewhere at a public place

Universal service

- ✓ Possibility for each individual or household to have the service, either privately or at home, or carried more and more to the individual's home by wireless devices

Objectives pursued by universal access and service

3 key objectives

- ✓ Availability
- ✓ Accessibility
- ✓ Affordability

Digital divide

- ✓ Disparity in ICTs access, use and competence
- ✓ Disparity between the haves and the have-nots in ICTs
- ✓ Abundance of ICTs means (terminals, connections, skills) for a group of people, on the one hand, and lack of means for another group of peo-

66 The four pillars of the telecoms http://www.telesperience.com/blog/the-four-main-pillars-of-the-telecoms-customer-experience

ple, on the other hand

- ✓ Several means of communication for the haves
  - o Access to information (radio and television)
  - o Telephone line and cell phones
  - o High speed internet access
- ✓ Only traditional means only for the have-nots (radio or television stations)

Types of digital devide

- ✓ Digital divide between two continents
- ✓ Digital divide between two countries
- ✓ Digital divide within the same country

Digital divide categories

- ✓ *First degree digital divide*: Unequal access
  - o No means of access (terminals, connections)
- ✓ *Second degree digital divide*: Unequal use
  - o Absence of offer of ICTs-related services
- ✓ *3rd degree digital divide*: Inequality of competence
  - o Lack of competence of any kind for the operation of telecommunications services

Causes of the digital divide

- ✓ Unavailability of basic telecommunications infrastructure in some regions
- ✓ Unavailability of broadband telecommunications infrastructure in some regions
- ✓ Inadequate ICTs democratization policy
- ✓ Low purchasing power of consumers
- ✓ Lack of ICTs training
- ✓ Lack of awareness on ICTs benefits

Most common mistakes in the use of telecommunications services

The most common mistakes in the use of ICTs
ICTs services: compliance with the principles and technical rules

1.- Forgotten password

Mistake committed most often

- ✓ *Password*: second tool after username for accessing Internet services
- ✓ *Password*: requirement of each account opened

Causes of forgotten password

- ✓ Multiplicity of passwords
- ✓ Complication of the password
- ✓ Infrequent use

Solutions

- ✓ Definition of 2 or 3 passwords to use for all accounts

2.-Choice of a weak password

- ✓ Password easily imaginable or detectable
- ✓ Example of a weak password
  - o 1234
  - o password
  - o birth date
  - o first name of a son or wife

Consequences of a weak password

- ✓ Unauthorized access to other persons to personal contents (e-mail, social networks, online banking, mobile banking, etc.)

Solution

- o Choice of a strong password (complicated password)
- o Techniques: Combination of letters, numbers, special characters (@, #, etc.), upper case, lower case

3.-No disconnection to the Internet

- ✓ Common mistake after an Internet session
- ✓ Open account = access to other people
- ✓ Open account: irreparable consequences

Solution

- ✓ Disconnection after each session regardless the terminal used (computer, tablet, cellular telephone, etc.)

4.-Computer always on

- ✓ Situation caused by different circumstances (urgent need for travel, omission, etc.)

- ✓ Open computer: access to unauthorized persons

Consequences

- ✓ Modification of documents
- ✓ Deletion of files
- ✓ Theft of documents
- ✓ Use of personal accounts by a third party

Solutions

- ✓ Shutdown of the computer with each move

5.-**Forgotten Attachments**

- ✓ Uncommon mistake
- ✓ Attachments (attached files): Important documents accompanying an e-mail
- ✓ Key part of the message

Solution

- ✓ Loading of the attachments just after entering the recipient's addresses and before writing the cover message

6.- **Reply to all by mistake**

- ✓ Frequent mistake in administrative exchanges

Consequences

- ✓ Answer sent not related to all recipiants copied in the message

Solution

- ✓ Deletion of addresses of recipients not targeted by the reply before writing the message

7.-**Unspecified file attachment**

- ✓ Common mistake in emails with attachments
- ✓ Mistake: choice of another file or folder instead of the one specified

Consequences

- ✓ Disturbance or nuisance for the sender and the recipient

Solution

- ✓ Validation of the name of the file before sending (validation of the new version of the modified files)

8.-**Display of an unspecified image**

- ✓ Common mistake in social networks
- ✓ Unspecified Image not shown
- ✓ Viewing of intimate photos or private images

Consequences

- ✓ Nuisance to the personal image of the sender

Solution

- ✓ Appropriate names for the photos and validation before any display

9.-**Sending of message to the unspecified recipient**

- ✓ Mistake most often in emails, sometimes on social networks

Consequences

- ✓ Nuisance for the sender, delayed waiting for the indicated recipient, etc.

Solution

- ✓ Correct entry of the recipient's email address or double check in the case of a choice in a contact list

10.-**Downloading of any file**

- ✓ Mistake occurred in emails, social networks, and searches

Consequences

- ✓ Virus infection, file loss, computer formatting, etc.

Solutions

- ✓ Downloading of messages from trusted contacts
- ✓ Attention to computer alerts during downloading to stop the process in time

11.-**Negligence of the folder of unwanted emails (Spam)**

- ✓ Mistake occurring in the use of emails
- ✓ Unwanted mail (junk mail, junk mail, spam): folder containing suspicious messages by a filter
- ✓ Important and authentic messages sent by the filter to the spam folder by mistake

Consequences

- ✓ Loss of important messages (nuisance and loss of opportunities related to this message)

Solutions

- ✓ Consultation at least 2 times a day the spam folder and transfer of authentic messages to the inbox (reporting the email address of the message as non-junk for future times)

List of core ICTs indicators[67]

Fundamental Indicators on Access and Infrastructure

Short list of core indicators

- ✓ Fixed telephone lines per 100 inhabitants
- ✓ Subscribers to cellular mobile services per 100 inhabitants
- ✓ Computers per 100 inhabitants
- ✓ Internet subscribers per 100 inhabitants
- ✓ Subscribers to a broadband Internet service per 100 inhabitants
- ✓ International Internet bandwidth per capita
- ✓ Percentage of population covered by mobile cellular
- ✓ Internet access rates (20 hours per month), in USD, as a percentage of revenue per person
- ✓ Cellular mobile telephone rates (100 min usage per month), in USD, as a percentage of revenue per person
- ✓ Percentage of localities (rural / urban) with public Internet access centers, by number of inhabitants

Extensive list of core indicators

- ✓ Radio receivers per 100 inhabitants
- ✓ TV receivers per 100 inhabitants

Core Indicators on ICTs Access and Use by Households and Individuals

Short list of core indicators

- ✓ Proportion of households having a radio receiver
- ✓ Proportion of households having a television receiver
- ✓ Proportion of households having a fixed telephone line
- ✓ Proportion of households having a cellular mobile telephone
- ✓ Proportion of households having a computer
- ✓ Proportion of people having used a computer (all connection locations combined) in the last 12 months
- ✓ Proportion of households having a home Internet access
- ✓ Proportion of people having used the Internet (all connection locations combined) in the last 12 months
- ✓ Location of Internet use by individuals in the last 12 months:
  - o Home
  - o Workplace
  - o Place of study
  - o Home of another individual
  - o Free Internet Access Facility;
  - o Commercial Internet Access Facility;
  - o Other
- ✓ Internet-related activities undertaken by individuals in the last 12 months
  - o For information:
    - Regarding goods or services;
    - Regarding health or health services;
    - From government organizations or public authorities via websites or e-mails;
    - Other or general web browsing
  - o For communication
  - o Purchase or order of goods or services
  - o Banking or other financial services
  - o Educational activities
  - o Relations with governmental organizations or public authorities
  - o Hobbies:
    - Downloading/practice of video or electronic games;
    - Acquisition of films, music or software;

67 INDICATEURS FONDAMENTAUX RELATIFS AUX TIC https://www.itu.int/en/ITU-D/Statistics/Documents/partnership/CoreICTSSSIndicators_f.pdf

- Reading/downloading of books, newspapers or magazines online;
- Other recreational activities

Extensive list of core indicators

- ✓ Proportion of people using a mobile telephone
- ✓ Proportion of households having access to the Internet, by type of access from home
- ✓ Frequency of individual access to the Internet in the last 12 months (all connection locations combined):
  - o At least once a day;
  - o At least once a week but not every day;
  - o At least once a month but not every week
  - o Less than once a month

Core indicators on access and use of ICTs by enterprises

Short list of core indicators

- ✓ Proportion of companies using computers
- ✓ Proportion of employees using computers
- ✓ Proportion of companies using the Internet
- ✓ Proportion of employees using the Internet
- ✓ Proportion of companies on the Web
- ✓ Proportion of companies using an Intranet
- ✓ Proportion of companies receiving orders over the Internet
- ✓ Proportion of companies placing orders via the Internet

Extensive list of core indicators

- ✓ Proportion of companies with access to the Internet by modes of access
- ✓ Proportion of companies having a local area network (LAN)
- ✓ Proportion of companies having an extranet
- ✓ Proportion of companies using the Internet by type of activity
  - o Reception and sending of e-mails
  - o For information:
    - On goods or services;
    - From government organizations or public authorities, via websites or e-mails;
    - Other information research or research activities
  - o Execution of banking transactions or access to other financial services
  - o Relations with governmental organizations or public authorities
  - o Provision of customer services
  - o Online sales of products

Core Indicators for the ICTs Sector

Short list of core ICTs indicators

- ✓ Proportion of the active labor force in the business sector in the ICTs sector
- ✓ Value added in the ICTs sector (expressed as a percentage of the total value added of the business sector)
- ✓ Imports of ICTs goods as a percentage of total imports
- ✓ Exports of ICTs goods as a percentage of total exports

Consumer rights

Consumer protection related to the following elements

- ✓ Telecommunications product
- ✓ Telecommunications equipment
- ✓ Telecommunications terminals
- ✓ Telecommunications Services

Main consumer rights

- ✓ Right to satisfaction of basic needs
- ✓ Right to a warranty on the product
- ✓ Right to information
- ✓ Right to choose
- ✓ Right to representation
- ✓ Right to appeal
- ✓ Right to education

- ✓ Right to a healthy environment[68]
- ✓ Right to cancellation and modification of the contract
- ✓ Compensation in case of service interruption
- ✓ Right to block advertising
- ✓ Access to emergency numbers
- ✓ Right to portability of the phone number

Main types of disputes between consumers and service providers[69]

- ✓ BBilling
  - o Package Migration
  - o RRebates
  - o Increase in price
  - o Application of specific tariffs
- ✓ Contracts
  - o Compliance with the implementing rules
  - o Compliance with the general conditions of subscription
- ✓ Technical issues
  - o Unavailability of the service
  - o Service operation problem
- ✓ Termination
  - o Contract cancellation difficulties
  - o Termination Fee

QUALITY OF TELECOMMUNICATIONS SERVICES

Some concepts

- ✓ Quality of service (QoS, quality of service): The set of characteristics of a telecommunication service enabling it to meet the explicit needs and the implicit needs of the service user.
- ✓ Quality of service requested by the user / customer (QoSR, QoS requirements): Quality of service requested by a customer / user or by one or more segments of the population of customers / users with the same performance requirements.
- ✓ Quality of service offered / planned by the service provider (QoSO, QoS offered): Level of quality expected and therefore offered to the customer by the service provider
- ✓ Quality of service delivered / obtained by the service provider (QoSD, QoS delivered): Level of quality of service obtained or delivered to the client
- ✓ Quality of service perceived or experienced by the client / user (QoSE, QoS experienced): Quality level used / actually operated by customers / users[70]

Quality Basis of Telecommunications Services

- ✓ Quality of service based on different parameters or characteristics
- ✓ QoS based on criteria defined by organizations
- ✓ QoS based on user perception of services

Elements of quality of service

- ✓ Availability of information transfer means
  - o Possibility for equipment and links to break down
- ✓ Maximum error rate
  - o Number of errored / modified bits in relation to the number of bits transmitted
- ✓ Transfer rate
  - o Transmission speed of information of access network
- ✓ Congestion
  - o Clutter of backbone networks
- ✓ Latency

68 Droit du consommateur des Telecom et TIC http://actic.over-blog.com/pages/Droits_Du_Consommateur_Des_Telecoms_Et_TIC-2421140.htmll

69 Résolution des litiges en 2015 http://www.mediateur-telecom.fr/ressources/media/files/AMCE_2015_bd.pdf

70 Série e: exploitation générale du réseau, service téléphonique, exploitation des services et facteurs humains https://www.itu.int/rec/dologin_pub.asp?lang=e&id=T-REC-E.800...F....

- o Transmission delay
- ✓ Jitter
  - o Variation in processing time
- ✓ Time delay
  - o Duration between the decision to issue the information and the receipt by the recipient[71]
- ✓ Reliability
  - o Availability of the service permanently
- ✓ Loss
  - o Loss of information during transmission

Quality indicators for telecommunications services

- ✓ A telephone call: ability to place a call with ease in less than 30 seconds
- ✓ Successful voice communication: A telephone call and its maintaining for 2 minutes without interruption
- ✓ Hearing quality: Hearing quality scored on a 4-level scale "perfect, acceptable, poor, and bad"
- ✓ Access to the internet: access to a page of a website within less than 10 seconds
- ✓ Broadcasting of video streaming: Viewing of a 2-minute video on a smartphone and check of the user's ability to access content normally

Sensitivity of the user to the following parameters:

- ✓ Transmission delay
- ✓ Variation of the transmission delay
- ✓ Loss of information

Quality of Service criteria defined by ITU

- ✓ Availability of the link (link connecting the two final consumers)
- ✓ Number of binary errors
- ✓ Transfer time
- ✓ Variation in the transfer delay

71 Réseaux et Télécommunications - Dominique SERET, Ahmed MEHAOUA, Neilze DORTA http://www.mi.parisdescartes.fr/~mea/cours/L3/L3.poly06.pdf

Quality of service from the point of view of the user and the service provider

- ✓ Quality of service requested by the user
- ✓ Quality of service provided by the service provider
- ✓ Quality of service delivered by the service provider
- ✓ Quality of service perceived by the user[72]

Levels of service

3 levels of service

- ✓ Best effort (no effort or lack of QoS): no differentiation between multiple network bitrate and *no guarantee.*
- ✓ Differentiated service or soft QoS: Definition of priority levels for different network bitrate without providing a strict guarantee
- ✓ Guaranteed service (hard QoS): reservation of network resources for certain types of flows.[73]

Obligations incumbent on any operator with regard to quality of service

- ✓ Providing permanent and continuous operation of the network and communication services
- ✓ Addressing the effects of system failure degrading quality of service
- ✓ Implementing the necessary equipment and procedures
- ✓ Measuring the quality of service indicators
- ✓ Complying with international standards (ETSI, ITU-T, AFNOR, ITIL, ITSEL, CENISSS, etc.)[74]

Obligations for the fixed service

- ✓ Permanence of the service
- ✓ Cumulative duration of unavailability
- ✓ Loss rate of internal network communications

Obligation for the mobile service

- ✓ Implementation of means to achieve service levels comparable to international standards

72 http://www.fratel.org/wp-content/uploads/2013/10/Pr%C3%A9sentation-Gide-R%C3%A9my-Fekete.pdf

73 http://www.fratel.org/wp-content/uploads/2013/10/Pr%C3%A9sentation-Gide-R%C3%A9my-Fekete.pdf

74 La qualité de service : quel rôle du régulateur pour quels objectifs ? http://www.fratel.org/wp-content/uploads/2013/10/Pr%C3%A9sentation-Gide-R%C3%A9my-Fekete.pdf

- ✓ Compliance with minimum conditions (call blocking rate, call cutoff rate, field strength and hearing quality)[75]

List of quality of service criteria

1. Bit rate (or bandwidth): maximum amount of information per unit of time
2. Transmission delay (or latency): delay between sending and receiving a packet
3. Variation of the transmission delay (or jitter): fluctuation of the digital signal, in time or in phase
4. Loss of package: non-delivery of a data packet, mostly due to network congestion
5. Inequencing: change of order of arrival of packets[76]

Quality of Service for Mobile Networks

- ✓ Quality criteria
- ✓ Network quality
- ✓ Terminal performance
- ✓ Type of contract (plan)
- ✓ Location of the user (inside or outside a building)
- ✓ Position relative to base stations
- ✓ Number of users connected simultaneously (data variable any time)[77]

Interactive and real-time services (telephony, video telephony, video games in real time)

Rigorous Services in terms of time constraints

- ✓ Transfer delay or signal transmission delay from source to destination
- ✓ Transfer delay or signal transmission delay: processing time + transmission and propagation time
- ✓ Transfer delay or signal delay: low and severe
- ✓ Variation in transfer delay or jitter
- ✓ Telephony and videoconferencing service: tolerance of a certain error (but no delay)

Quality constraints for data services

- ✓ Bit rates: minimum bit rate to be guaranteed, peak bit rate, average bit rate
- ✓ Errors: binary error rate (number of errored bits out of transmitted bits), duplication or insertion of packets, order of packets, etc.
- ✓ Data service: tolerance of a certain transmission delay (but no binary error)

Constraint for multimedia service type

- ✓ Need for the network to support several types of services with different constraints[78]

SOME TELECOMS SERVICES

Fax (Facsimile)

- ✓ Transmission of graphic documents by electrical or electronic means via a telephone line
- ✓ Techniques for remote reproduction of graphic documents by means of terminals (fax machines) connected to the telephone network (access conditioned by the composition of a similar fax number in format and operation with a telephone number)
- ✓ Reproduction identical to the original, usually in black and white, sometimes in shades of gray, or in color
- ✓ Electronic document digitized by a machine and sent electronically to another printing machine

Fax machine

- ✓ Electronic device used to send documents electronically over a telephone network.

Principles of Operations of fax

- ✓ Insertion of the document to be transmitted in the fax machine
- ✓ Dialing of the receiving fax number (format of a regular telephone number)
- ✓ Validation of the sending by pressing the "*send*"

---

75 La qualité de service : quel rôle du régulateur pour quels objectifs ? http://www.fratel.org/wp-content/uploads/2013/10/Pr%C3%A9sentation-Gide-R%C3%A9my-Fekete.pdf

76 La qualité de service : quel rôle du régulateur pour quels objectifs ? – http://www.fratel.org/wp-content/uploads/2013/10/Pr%C3%A9sentation-Gide-R%C3%A9my-Fekete.pdf

77 Qualité des services de communications électroniques – www.telecom-infoconso.fr/qualite-des-services-de-communications-electroniques/?

78 Réseaux publics de télécommunications (Ed. 3.4. Revision : 9/01) – http://www.ulb.ac.be/students/bep/files/intro3.4.pdf

button to send the fax

- ✓ Line by line scan of the physical document inserted for transmission by the fax scanner
- ✓ Conversion of the document image into electrical pulses (electrical signals)
- ✓ Transmission of electrical pulses via an ordinary telephone line or a leased line
- ✓ Reception of electrical signals by the destination fax machine
- ✓ Conversion by the receiving fax machine of electrical pulses into images
- ✓ Transcription of images on paper

Classification of fax machines

4 groups of fax machines

- ✓ *Group 1:* low resolution analog fax machines (nonexistent today)
  - o Transmission by frequency modulation
  - o Six minutes for transmission of an A4 page
- ✓ *Group 2:* Low Resolution Analog Faxes
- ✓ Transmission by amplitude modulation
- ✓ *Group 3:* "digital" fax machines on a good-resolution analogue network (200 × 196 dots per inch)
- ✓ *Group 4:* "digital" fax machines using the ISDN network at 64 kbit /s
  - o Digital photocopy quality (400 ×400 points per inch[79]

Internet Fax (e-fax, online fax)

- ✓ Techniques allowing the use of IP networks for document transmission
- ✓ Sending of digital documents to a fax machine from a computer or digital terminal using a software
- ✓ Replacement of all end-to-end transmission infrastructures with fax software

Internet Fax Services

- ✓ Email to fax (fax from an existing email application)
- ✓ Fax to email (receiving faxes in the mailbox)
- ✓ Sending of e-mails with attachments to traditional fax machines (connected by telephone line or connected to the telephone network) over a telephone line
- ✓ Conversion of faxes received from traditional fax machines to e-mail
- ✓ Sending of digitized documents to e-mails
- ✓ Computer to fax (Sending of a fax from a computer)

Email

Email (electronic mail), or email address

- ✓ Service of transmission of written messages accompanied by attachments (text, sound, music, image, video, data)
- ✓ Transmission of multimedia messages through a computer network, mainly the Internet
- ✓ *E-mail*: term used to designate both the e-mail address and the electronic message sent by this means
- ✓ First e-mail sent in the Autumn 1971 by Ray Tomlinson, inventor of e-mail

Constitution of an email address

5 elements

1. Username or account name or login (Jean, Pierre, Simone, contact, info)
2. @: Symbol used to separate the user name from the domain (read at sign, at)
3. Domain (Hosting, Hosting Server, Mail Server)
4. **.** (dot): used to separate the host from its extension
5. Domain extensions (com, fr, org, int, edu)
   - E-mail: dupont@domain1.fr
   - E-mail: gregorydomond@hotmail.com
   - E-mail: itumail@itu.int

Parts of an E-mail (electronic message)

2 parts

1.- Header

- ✓ Subject of the message
- ✓ Sender's name
- ✓ Sender's email address

79 Télécopieur https://fr.wikipedia.org/wiki/Telecopieur

- ✓ Date and time of reception
- ✓ Answer (with "reply to" possibilities)
- ✓ Recipient's name
- ✓ Recipient's email address

2.- Content

- ✓ Body of the message (content)
- ✓ Possible attachments
- ✓ Signature of the sender

Principle of operation of the email

- ✓ An email address for the sender of the message (opening an account with an email service provider or a professional account with an employer)
- ✓ E-mail addresses of the recipients of the message
- ✓ Access to the account by the user (sender) using the secret and secure password
- ✓ Entry of the recipient's email address (or recipient's email addresses) in the reserved space
- ✓ Entry of the subject of the message in the reserved space
- ✓ Entry of the message in the reserved space (possibility to copy and paste text taken elsewhere)
- ✓ Possible addition of attachments (text, sound, music, video, data)
- ✓ Press of the "send" button to transmit the message
- ✓ Transit of the message from servers to servers to the destination server
- ✓ Delivery of the message in the recipient's mailbox
- ✓ Mailbox: a reserved space in a mail server for each user (account holder at a service provider)
- ✓ Access for the recipient to the message by entering his/her username and password
- ✓ Reading of the message and downloading of any documents attached to the message on his or her terminal (computer, digital tablet, smartphone)

Features of an email address

- ✓ Electronic mailbox for each user
- ✓ Notification of the recipient of the arrival of each message
- ✓ Confirmation of receipt of the message by an acknowledgment of receipt
- ✓ Original message included in answer
- ✓ Sending of the same message to multiple recipients at the same time
- ✓ Ability to retrieve deleted messages for a while[80]
- ✓ Ability to save contact email addresses in an account directory
- ✓ Possibility of simultaneous sending of message to the main recipients, the recipients in carbon copy (Cc: carbon copy) and recipients in blind copy (Bcc or Ccc)
- ✓ Notification by the mail server of the non-delivery of the message in the event of an erroneous address

Access to the email service

- ✓ Computer or digital tablet or cellular telephone connected to a telecommunications network
- ✓ Access to an email server (ISP or public mail server such as: Hotmail. Gmail, Yahoo, etc.)
- ✓ email address
- ✓ Messaging software installed on the terminal (in some cases)

Advantages of email

- ✓ Speed of message circulation
- ✓ Ability to send messages to non-connected recipients
- ✓ Timely viewing of messages
- ✓ Abolition of Distance
- ✓ Ease of use
- ✓ Economy (compared to traditional mail)
- ✓ Technical universality (use of the service on all types of equipment, networks, etc.)
- ✓ Group communication (possibility of group sending, mailing list, distribution of messages in newsgroups and forums)
- ✓ Written trace of messages (data archiving, data

80 La messagerie électronique http://hautrive.developpez.com/reseaux/?page=page_20

mining, message classification)

- ✓ Reuse of messages (reply, message transfer, etc.)[81]

Disadvantages of email

- ✓ Reduced message privacy (ability to intercept messages during their transmission across networks)
- ✓ Propagation of viruses (main vector for spreading computer viruses)
- ✓ Abusive messages (unsolicited mail, spam, etc.)
- ✓ Propagation of rumors (hoax)
- ✓ Information overload or informational deluge (large volume of e-mails to be processed by a user during a day)[82]

Cellular network messaging services

SMS and MMS

SMS : Short Message Service

- ✓ Text messaging service provided by mobile networks
- ✓ Content of the message: 160 alphanumeric characters
- ✓ Service *Store and Forward*: Message stored first in the servers, and forwarded to the recipient
- ✓ Network used: from 2 G
- ✓ SMS delivery
  - o almost instant
  - o after a while (at peak times or busy hours)
  - o message not delivered due to some technical constraints

MMS Multimedia Messaging Service : Multimedia messaging service

- ✓ Sending and reception of multimedia messages by mobile telephone
- ✓ MMS: Extension of the capabilities of an SMS
- ✓ Size: between 300 and 600 KB
- ✓ Service *store and forward (*Message stored first in the servers, and forwarded to the recipient)
- ✓ Network used: from 2.5 G
- ✓ Delivery of the MMS
  - o almost instant
  - o after a while (at peak times or busy hours)
  - o message not delivered due to some technical constraints

Instant messaging

- ✓ Online chat
- ✓ Exchange of text messages or files (images, video, sound, ...) in real time between several users connected to the same network (computer network or Internet)
- ✓ Internet software or application allowing real-time communication between connected users (online)
- ✓ Software or Internet application to talk to each other using a microphone, to see each other by webcam, and to share multimedia files in real time
- ✓ Service to visualize in real time the presence and availability of contacts
- ✓ Exchange possible only with people in the contact list (different from chat: live exchange with any stranger)

Types of instant messaging

2 types of instant messaging

Fixed instant messaging

- ✓ Computer-based operation

Mobile Instant Messaging

- ✓ Cellular telephone Based Operation

Principle of operation of instant messaging

- ✓ Network connection (physical connection to the network, usually the Internet)
- ✓ Connection to the server containing information about registered users, connected or not (Client software for connection with the instant messaging server)
- ✓ Notification of the connected user of the online presence of contacts

81 Principes de fonctionnement de la messagerie électronique https://www.sites.univ-rennes2.fr/urfist/messagerie_electronique_fonctionnement

82 Principes de fonctionnement de la messagerie électronique https://www.sites.univ-rennes2.fr/urfist/messagerie_electronique_fonctionnement

- ✓ Transmission of messages via the mail provider server
- ✓ Connection of the messaging with the server at each launch
- ✓ Querying of the database through email to see connected users
- ✓ Possible exchange of messages between connected people

Telegraphy

- ✓ System for transmitting messages called telegrams over long distances using codes

Types of telegraphy

- ✓ Optical telegraphy
- ✓ Electric telegraphy
- ✓ Wireless telegraphy

Optical telegraph of Chappe

- ✓ Transmission of signals (coded messages) step by step thanks to a network of semaphores

Electric telegraph

- ✓ Transmission of signals via electrical wires day and night and whatever the weather conditions
- ✓ Establishment of intercontinental links using submarine cables

Principle of operations of Optical Telegraphy (Aerial Telegraphy)

- ✓ Exchange based on stations (Chappe towers built on high points) in direct view of each other
- ✓ Installation of a machine in each tower
- ✓ A large arm and several small wooden arms in each machine
- ✓ Swiveling of arms to take hundreds of distinct positions
- ✓ Indication of a specific message by each of the positions
- ✓ Transmission of messages as a code
- ✓ Communication achieved by arms supported by a mobile mast of the machine
- ✓ Reproduction of signals emitted by the arms
- ✓ Use of the same codes on transmission and reception
- ✓ Service not usable during night, nor by poor visibility

Principles of operations of electric telegraphy

- ✓ Use of batteries, a switch, an electromagnet and two wires to transmit the two signs: a short and a long one (a dot and a line)
- ✓ Use of the same codes on transmission and reception

Principles of operations of wireless telegraphy

- ✓ Transmission of information using electromagnetic waves
- ✓ Using the Morse Alphabet for encoding information to be transmitted
- ✓ Use of the same codes on transmission and reception

Telex

- ✓ Telegraph service enabling its subscribers to correspond directly and temporarily with each other by means of start-stop device and telecommunication circuits of the public telegraph network[83]
- ✓ International communication network linking teletypewriter (terminal devices for sending and receiving messages by means of electrical signals)
- ✓ Teletype communication network
- ✓ Service derived from telegraphy

Principles of operation of the telex

- ✓ Switched teletypewriter network for the sending of text messages
- ✓ Routing of text messages by using a telex address (for example: 15710 GD H, 15710: User number, GD: abbreviation of the user name and H: Country code)
- ✓ Possibility of routing messages to different telex terminals in the same company using the identities of the different terminals (for example: T150, T151, etc.)
- ✓ Acknowledgment of receipt of the telex message

83 Recommendation UIT- R v.662-3
Termes et définitions
https://www.itu.int/rec/R-REC-V.662/fr

by the message "Answerback"

Videotex (interactive videography)

- ✓ Videography service supported by a telecommunications network providing the transmission of user requests and messages obtained in response
- ✓ Access to information through dialogue with a data bank or terminal
- ✓ Service allowing the sending of pages of text and graphics to a user in response to his or her request
- ✓ Information system based on telecommunications, computer and television technologies

Types of Videotex

2 types of videotex

1.- Broadcast videotext or teletext

- o Computer system allowing the user to select and display information pages on the screen of a television receiver
- o Unilateral message (without interaction with the receiver)

Principles of operation of the broadcast videotext or teletext

- ✓ Pages transmitted by television or cable
- ✓ Use of television carrier waves to interpose information page signals between those of television program images
- ✓ Display information pages on the TV screen through a decoder (Access to information pages via a decoder)

2.- Interactive videotext or switched videotex (teletel system)

- o System for selecting and displaying information pages on a television receiver screen
- o Bidirectional computer system for obtaining information or conducting transactions
- o Selection of information by dialogue with the central computer of the system
- o Selection and display of information pages for consultation

Principles of operation of interactive or switched videotex

- ✓ Connection of the subscriber to the teletel center using an alphanumeric keypad (Minitel terminal)
- ✓ Request of a service by dialing the number on the keypad
- ✓ Establishment by the center of a connection with the computer managing the requested service
- ✓ Start of the dialogue between the user and the computer managing the service

Uses of Interactive Videotex

- o Orders from service providers
- o Booking of tickets
- o Sending of e-mail

Video Calling

- ✓ Association of telephony and television allowing users to see each other during their telephone conversation[84]
- ✓ Service supported by fixed or mobile telephone networks or the Internet

Principles of operations

- ✓ Coding and decoding of analog audio and video signals into digital signals at both ends using a Codec (Encoder and Decoder)
- ✓ Compression of signals digitized by the codec
- ✓ Transmission of signals via the telephone network
- ✓ Decompression and conversion of digital signals into analog signals on reception

Videophone service minimal requirements

- ✓ 3G mobile telephones at both ends (best quality with a 4G terminal)
- ✓ 3G mobile telephone network (UMTS) (Best quality with a 4G / LTE network)

Videoconference

- ✓ Video teleconference or interactive television
- ✓ A set of interactive telecommunication technologies allowing two or more remote sites to interact with audio and video transmissions in simultaneously directions
- ✓ Simultaneous transmission of audio and video information between two or more remote locations.
- ✓ Telephone conversation accompanied by images

84 Visiophonie https://tutovideo.ch/actualites/tooltip/videophonie/

between two or more people in two public or private administrations, or by a multi-site videoconference between several people in a meeting room

Components of a videoconferencing system

Main components at both ends

- ✓ *Codec (Encoders and decoders)*: Devices for conversion of signals into compressed digital form prior to transmission
- ✓ *Cameras*: Capture and conversion of images into electrical signals
- ✓ *Screens*: Display of images
- ✓ *Microphones*: Capture and conversion of the sound into an electrical signal
- ✓ *Speakers*: Reception of the electrical signal and conversion into an audio message
- ✓ *Video projector*: Device projecting images from a computer
- ✓ *Transmission system*: Establishment of a private transmission system or connection to the ISDN or IP network

Phases of a videoconference

- ✓ Communication of a conference telephone number to all participants
- ✓ Call to the number indicated at the indicated time
- ✓ Automatic connection of participants after the call (virtually in the meeting room: appearance of the participant on the screen of the other conference room)
- ✓ Start of videoconference after the connection of all participants (virtually in the meeting room)
- ✓ Exchange of sounds and images in real time

Audio conference or conference call

- ✓ Teleconference service based on the connection of several participants by telephone circuits (with possible addition of the transmission of signals such as faxing or telewriting)
- ✓ Service allowing multiple people in different locations to communicate by telephone in real time

Principles of operation of the audioconference

- ✓ Establishment of a conference bridge (equipment connecting telephone lines)
- ✓ Connection of each participant to the bridge conference by dialing a telephone number
- ✓ Start of conversation after the login of all participants

CHAPTER 4

TELECOMMUNICATIONS TECHNIQUES AND TECHNOLOGIES

MAIN OBJECTIVE OF TELECOMMUNICATIONS TECHNIQUES AND TECHNOLOGIES

- ✓ Transmitting information from end to end in electrical, electromagnetic and light forms

Specific objectives of telecommunications techniques and technologies

- ✓ Acquisition or capture of information from any point
- ✓ Conversion of information in other forms
- ✓ Processing of information (transmit side)
- ✓ Transmission of information
- ✓ Processing of information (receive side)
- ✓ Conversion of information into original form

Strategies used in telecommunications

- ✓ Electronic terminal devices for acquiring information
- ✓ Electronic devices for information processing
- ✓ Transmission media for transmission of information
- ✓ Electronic terminal devices for the use of information

Basic telecommunications principles

- ✓ Physical and electromagnetic laws
- ✓ Electrical and mechanical laws
- ✓ Electronic principles
- ✓ Principle of radio conduction
- ✓ Principle of semi-conductors
- ✓ Propagation of electromagnetic waves
- ✓ Analog technologies
- ✓ Digital technologies
- ✓ Optics

Technical foundations of telecommunications

Techniques supported by

- ✓ Laws of Physics (propagation in guided and unguided environments)
- ✓ Circuit Theory
- ✓ Theory of electromagnetism
- ✓ Signal processing (coding, modulation, detection,)
- ✓ Electronics and optoelectronics (manufacturing of devices)
- ✓ Telecommunications links

TELECOMMUNICATIONS TECHNIQUES

- ✓ Transmission techniques (remote transmission of information)
- ✓ Switching techniques of any two users according to their orders (*switching*)[85]

Basis of telecommunications services

- ✓ Techniques
- ✓ Technologies
- ✓ Terminal equipments
- ✓ Infrastructures
- ✓ Links
- ✓ Telecommunications Resources
- ✓ Standardization

Means used in telecommunications

- ✓ Electrical and electronic systems
- ✓ Propagation in guided media
- ✓ Propagation in unguided media (in free space)
- ✓ Techniques
- ✓ Technologies
- ✓ Systems and devices
- ✓ Communication channels

Solutions to problems of remote transmission of information by technical means

Based on the following factors

- ✓ Speed of propagation of electromagnetic phenomena

85 Introduction aux télécommunications http://www.volle.com/ENSPTT/introtcom.htm

- ✓ Propagation without physical medium
- ✓ Propagation via physical medium
- ✓ Easy conversion of physical quantities into electrical signals
- ✓ Speed of execution of electronic devices
- ✓ Extreme variety of achievable electronic functions
- ✓ Miniaturization offered by microelectronic technologies
- ✓ Arrival of transmission means (optical fibers), coherent sources (lasers), modulators and detectors working efficiently at optical frequencies[86]

Steps of Telecommunications

1- *Acquisition of information*
   - o Use of a terminal for the capture or acquisition of information
   - o Transduction of information into electrical signal

2- Signal processing
   - o Signal transformation

3- Signal transmission
   - o Transfer of the signal from one point to another

4- Use of signal
   - o Use of a terminal for signal transduction in information
   - o Restitution of the original information

User information

- ✓ Speech
- ✓ Music
- ✓ Texts
- ✓ Still images
- ✓ Moving images
- ✓ Data[87]

Primary sources of information

- ✓ Human being: voice, image, gesture
- ✓ Machines: Control information, calculation results
- ✓ Information bank: Books, records, tapes, computer memories
- ✓ Natural environment : Physical phenomena observed[88]

Nature of the information transmitted

- ✓ Text (telegraphy, telex)
- ✓ Sounds (telephony and broadcasting)
- ✓ Images (television and facsimile)
- ✓ Data (teleinformatics)
- ✓ Measurements (radar, telemetry)
- ✓ Orders (remote controls, radio guidance)[89]

Information flow

- ✓ Dialogue between humans
- ✓ Dialogue between humans and machines
- ✓ Dialogue between machines
- ✓ Observation (measurement) of physical phenomena[90]

Information exchanged through a telecommunications system

- ✓ End user information
- ✓ Information for automatic systems
- ✓ Control information

Information and message

- ✓ *Information*: content of the message sent
- ✓ *Message*: Information envelope
- ✓ Message with information
- ✓ Message without information

86 Cours/Lecture series – 1986 -1987 Academic Training Programme – F. de COULON/EPF, Lausanne

87 Systèmes de télécommunications, Bases de transmission – P.-G. FONTOLLIET

88 Introduction à l'Electronique – http://www.epsic.ch/cours/electronique/techn99/elniq/ELNELT.html

89 A5 Systèmes électroniques http://www.epsic.ch/cours/electronique/techn99/elniq/elnsystxt.html

90 A1 Electronique - Electrotechnique http://www.epsic.ch/cours/electronique/techn99/elniq/ELNELT.html

Analog information
- ✓ Human voice (speech)
- ✓ Music
- ✓ Animated picture of television
- ✓ Light, temperature

Digital information
- ✓ Digital file
- ✓ Digital image
- ✓ Texts and data

Information transfer process
- ✓ Selection of messages by the source of information
- ✓ Transformation of the message by the transducer into an electrical or optical signal
- ✓ Various processings of the electrical signal by the transmitter
- ✓ Transfer of the electrical or optical signal via a communication channel
- ✓ Transformation of the electrical or optical signal into a message by the receiver
- ✓ Use of the message by the recipient[91]

Information Processing Tools
- ✓ Converter
- ✓ Filter
- ✓ Encoder
- ✓ Modulator

General structure of a chain of transmission (analog information)
- ✓ Source
- ✓ Transducer
- ✓ Transmitter
- ✓ Channel
- ✓ Receiver
- ✓ Transducer
- ✓ Destination

Information source
Information Production (Message)
- ✓ Sound
- ✓ Light
- ✓ Temperature
- ✓ Speed
- ✓ Acceleration
- ✓ Movement
- ✓ Strength

Transmit Transducer
Transformation of information into electrical signal
- ✓ Transmit Transducer: device converting the message from the source into an electrical signal
- ✓ Examples:
  - o Microphone
  - o Camera

Transmitter
Device for processing and transmitting electrical signals
- ✓ Preamplification
- ✓ Analog - Digital conversion
- ✓ Coding
- ✓ Modulation
- ✓ Filtering
- ✓ Power amplification

Channel
- ✓ *Path intended for transmitting the electrical signal from the transmitter to the receiver*
- ✓ *Physical or logical link connecting a data source to a data sink*
- ✓ *Example*
  - o Two-wire line
  - o Coaxial cable
  - o Optical Fiber

91 Cours/Lecture series – 1986 -1987 Academic Training Programme F. de COULON/EPF, Lausanne

o Free space

Receiver

Device for capturing and processing transmitted electrical signals

- ✓ Amplification of the received signal
- ✓ Filtering
- ✓ Demodulation
- ✓ Decoding
- ✓ Digital – Analog conversion
- ✓ Power amplification

Receive transducer

Transformation of electrical signals into initial information

- ✓ Receive transducer: device converting an electrical signal into an original message
- ✓ *Examples:*
  - o Loud speaker
  - o Display screen

Recipient

Reception and use of the transmitted message

Examples:

- o Human Users
- o Systems

General structure of a chain of transmission (digital information)

- ✓ Computer terminals or others
- ✓ DTE (Data Terminal Equipment) at the source
  - o Source or collector of data
  - o Communication controller
- ✓ Digital Interfaces (exchange between DTE and DCE)
- ✓ DTE (Data Circuit Termination Equipment = Modem)
- ✓ Analog interfaces (exchange between DCE and the transmission channel and vice versa)
- ✓ Channel (transmission line - baseband transmission)
- ✓ DTE (Data Circuit Termination Equipment = Modem)
- ✓ DTE (Data Terminal Equipment) at the destination[92]

Functional organization of the information channel

6 items

1. - Acquisition

- ✓ Capture and Transformation of a physical quantity into information : an electrical signal (analogical chain) or a set of numerical values (digital chain)[93]

Results of the acquisition

- ✓ Electrical signal
- ✓ Binary data

2.- Coding

Adaptation of the digital values to be transferred to the transmission channel

Coding steps

- ✓ Source coding: reduction of the size of the amount of information to be transmitted (data compression)
- ✓ Channel coding: protection of information to be transmitted or stored against errors
- ✓ Modulation: transformation of the binary values into an analog signal adapted to the transmission channel

Coding results

- ✓ Data adapted to the transmission channel

3.- Transmission - Storage

Transmission: Transfer of information from the source to the destination via a transmission channel

Storage: backup of the information on a physical medium

4.- Decoding- processing

- ✓ Achievement of reverse operations from those of "coding"

Demodulation

- ✓ Transformation of the electrical signal into a se-

92 Communications analogiques EII1 http://slidegur.com/doc/7603257/communications-analogiques-eii1-organisation-cm-td

93 Tronc commun - Approche fonctionnelle des systèmes

ries of binary symbols.

Channel decoding

- ✓ Detection and correction of possible errors

Source decoding

- ✓ Reconstitution of initial data by minimizing the compression gap

5.- Processing

- ✓ Transformation of initial data according to user's choice of presentation

6.- Restitution

- ✓ Setting the initial physical quantity available to the user in the desired format (transformation of the electrical signal or binary data stream into a physical quantity)[94]

Electronics and Information

- ✓ *Electronics*: Branch of Applied Physics, dealing with the shaping and management of electrical signals capable of transmitting or receiving information[95]
- ✓ *Electronics*: Set of techniques using electrical signals to capture, transmit and use information[96]
- ✓ *Electronics*: Branch of Applied Physics studying and designing structures processing electrical signals (electrical currents or voltages) carrying information[97]

Electronics: Telecommunications Basis

Inescapable base of telecommunications

- ✓ Tools for capturing or acquiring information
  - o Microphone, camera, keyboard and screen
- ✓ Information Processing Tools
  - o Filter, converter, amplifier, encoder, modulator
- ✓ Information transmission tools
  - o Movement of electrons in a conductor
  - o Propagation of radio waves in free space
- ✓ Information restitution tools
  - o Speaker, screen

Contribution of Electronics to Telecommunications

- ✓ Speed of propagation of electromagnetic phenomena
- ✓ Possibility of propagation in free space (transmission without physical medium)
- ✓ Conversion of physical quantities into electrical form
- ✓ Speed of execution of electronic devices
- ✓ Extreme variety of achievable electronic functions
- ✓ Miniaturization of microelectronic technologies[98]

CONTRIBUTIONS OF SCIENCE TO TELECOMMUNICATIONS

- ✓ Telecommunication development through various contributions from different scientific disciplines
- ✓ Use of 4 scientific domains: Mathematics, Physics, Software Engineering and Chemistry

Contribution of mathematics to telecommunications

- ✓ Significant Contributions of Mathematics to Telecommunication Development
- ✓ Description of telecommunications signals by mathematical equations
- ✓ Intervention of mathematical tools in signal processing, cryptography, information theory and digital technologies
- ✓ Cryptographic functions (confidentiality, integrity, authentication and non-repudiation) supported by mathematical tools
- ✓ Development of digital technologies by mathematical logics

---

94 Tronc commun - Approche fonctionnelle des systèmes

95 Chapitre 2 http://cedric.cnam.fr/~bouzefra/cours/Arduino1erePartie.pdf

96 A1 Electronique – Electrotechnique http://www.epsic.ch/cours/electronique/techn99/elniq/elnelttxt.html

97 L'Electronique http://electronique-web.blogspot.com/p/lelectronique.html

98 A1 Electronique – Electrotechnique http://www.epsic.ch/cours/electronique/techn99/elniq/elnelttxt.html

Contributions of physics to telecommunications

- ✓ Considerable contributions from physics to telecommunications
- ✓ Telecoms infrastructure supported by physical principles
- ✓ Use of radiocommunication (radio, television, cellular telephony, wireless internet, etc.) thanks to the theory of the propagation of electromagnetic waves
- ✓ Mobility of telecommunications based on physical principles
- ✓ Applications of the principles of basic Electronics for the acquisition, processing, transmission and use of information
- ✓ Essential roles of diodes, thyristors, integrated circuits and lasers made from semi-conductors
- ✓ Roles of Optoelectronics (Electronics and Photonics) dealing with emission and the reaction to light both essential to optoelectronic interfaces.

Contributions of computing to telecommunications

- ✓ Significant contributions of Computing to telecommunication development
- ✓ Dependence of Modern Telecommunications on Computing
- ✓ Software development adapted to the needs of telecommunications thanks to software engineering
- ✓ Many functions performed by software in telecommunications systems
- ✓ Roles of a software in the automatic processing of a telephone call (computer tools supporting the programming logic)
- ✓ Use of micro-computers both in telecommunications operations and for access to the Internet

Contributions of Chemistry to telecommunications

- ✓ Fundamental roles of chemistry in telecommunications systems
- ✓ Use of telecommunications services based on chemical principles
- ✓ Redox reaction for weight reduction and battery life of handheld devices[99]

Principles of Telecommunications

- ✓ Logical sequence guaranteeing
  - o acquisition of information
  - o processing of information
  - o transmission of information
  - o use of information
- ✓ Functioning of Telecommunications based on interactions between signals and electronic systems
- ✓ Electrical energy: "Language spoken" by telecommunications systems
- ✓ Transformation of the user's message into an electrical signal by interfaces
- ✓ Telecommunications system crossed end-to-end by the electrical signal
- ✓ Establishment of a physical or non-physical link between the transmitter and the receiver
- ✓ Frequency adjustment between the terminal and the system (the network)
- ✓ Common codes (same language) between transmitter and receiver for easy communication
- ✓ Infrastructure and embedded intelligence
- ✓ Equipment and connections needed

Principles of operation of telecommunication systems

Principles based on the interactions between signals and systems

- ✓ User signals (user messages transformed into electrical signals)
- ✓ Signals generated by systems in support of telecommunications activities
- ✓ Systems (transmitters and receivers, switches, servers, etc.)
- ✓ Signal Actions (Signal Controls) on Telecommunications Systems
- ✓ Answers provided by telecommunication systems

99 Télécommunications et sciences http://www.techno-science.net/?onglet=glossaire&definition=3982

Information exchange process between two users

- ✓ Production of the message (envelope of the information) to be transmitted by the source
- ✓ Conversion of the message into an electrical signal via an interface
- ✓ Generation of a Signal adapted to the transmission channel by the transmitter (adaptation of the signal to the transmission medium)
- ✓ Transmission of the signal through a transmission channel: link between transmitter and receiver
- ✓ Signal capture by the receiver
- ✓ Conversion of the electrical signal into a message
- ✓ Reception of the message by the recipient

SIGNALS AND SYSTEMS

2 essential elements in telecommunications operations

Signal

- ✓ Physical representation of information being transmitted
- ✓ Variable quantity over time, used to cause any effect or produce any action
- ✓ Physical media of information
- ✓ Means of conveying information
- ✓ Physical quantity varying over time and conveying information: voice, sound, music, image, video, text, data
- ✓ Function of time used to represent a variable of interest associated with a system
- ✓ Electric current, electromagnetic wave or light wave used to transmit information
- ✓ Entity (electric current, acoustic wave, light wave, sequence of numbers) generated by a physical phenomenon and carrying information (music, speech, sound, image, temperature)[100]

Signal examples

- ✓ Sound signals: fluctuations in air pressure carrying a message to the ear
- ✓ Visual signals: light waves bringing information to the eye[101]

Analog signals

Analog signal: continuous variation over time

- ✓ Information produced by the source having a variation or a continuous range of nuances[102]
- ✓ Naturally generated, continuous signals (sensors, amplifiers, DAC), processed by electronic circuits, (or manually)
- ✓ Infinity of different values in each range and continuously transmitting in the time axis[103]
  - o Examples: Word, Music, Image

Analog communication

- ✓ Transmission of information (voice, image, video, data) in the form of an analog wave

Digital signals

Digital signal: succession of discrete states

- ✓ Representation of the information produced by the source by a conventional system of distinct signs, or electrical quantities fixed in advance and limited to very few values (0V and 5V, for example)
- ✓ Signals used in computer processing
- ✓ Convenience and quick processing thanks to their artificial natures
  - o Examples: Binary sequence, computer file, Morse code, text

Digital communication

- ✓ Transmission of information (voice, data, images, etc.) in the form of bit
- ✓ Transmission of information (voice, image, video, data) in digital form, that is, using bits "1" and "0"

Signal processing

Process making the information converted into an electrical signal suitable for transmission

- ✓ Sets of steps for transforming the signal on transmission and reception

100 Traitement du signal - https://vdocuments.site/cours-traitement-signal-1.html

101 Systèmes de télécommunications, Bases de transmission - P.-G. FONTOLLIET

102 Le signal électronique - http://www.epsic.ch/cours/electronique/techn99/elniq/ELNSIGN.html

103 A1 ELECTRONIQUE - ELECTROTECHNIQUE - http://www.epsic.ch/cours/electronique/techn99/elniq/ELNELT.html

- ✓ Development of information-bearing signals
- ✓ Interpretation of information-bearing signals[104]

Signal processing Foundation

- ✓ Theory of signal and information
- ✓ Electronic Resources
- ✓ Computing Resources
- ✓ Applied Physics Resources[105]

Essential tasks of signal processing

- ✓ Extraction of useful information incorporated into signals
  - o Analysis, filtering, regeneration, measurement, detection, identification
- ✓ Representation of results in a form suitable for humans or machines
- ✓ Elaboration of signals allowing the study of the behavior of the physical systems
  - o Systems used to support the transmission or storage of information[106]

System

- ✓ Isolated set of devices establishing a cause and effect link between input signals (excitations, commands, directives, disturbances and output signals (responses or measurements)[107]
- ✓ Set of elements interacting with each other according to certain principles or rules[108]
- ✓ Organized collection of interacting objects to form a whole

Examples of telecommunications systems

- ✓ Radio and television station
- ✓ Telephone network
- ✓ Email
- ✓ Video conference

Basic components of a communication system

Main basic elements

- ✓ Communication Technologies
- ✓ Communication devices
- ✓ Communication channels
- ✓ Communication software

Digital communications process

- ✓ Conversion of analog information into digital signal (transmission)
- ✓ Transmission of digital signals as analog waves
- ✓ Digital signal conversion to analog information (reception)

Advantages of digital communication systems

- ✓ Immunity to noise
- ✓ Signal processing capacity
  - o Digital signals more suitable for processing
  - o Easier storage
  - o Bit rate adaptation
- ✓ Use of Repeater against noise
- ✓ Easy to measure and evaluate
- ✓ Detection and correction of transmission errors
- ✓ Implementation of flexible digital facilities
- ✓ Capacity to carry a combination of traffic (telephone signals, data, videos, teletext)

Disadvantages of digital communication systems

- ✓ Consumption of greater bandwidth (more frequency)
- ✓ Complexity of circuits (several analog - digital conversions and vice versa)
- ✓ Accurate synchronization (between transmitter and receiver clocks)[109]

Communication devices

- ✓ Hardware capable of transmitting data, instruc-

104 Cours/Lecture series – 1986 -1987 Academic Training Programme – F. de COULON/EPF, Lausanne

105 Cours/Lecture series – 1986 -1987 Academic Training Programme – F. de COULON/EPF, Lausanne

106 Cours/Lecture series – 1986 -1987 Academic Training Programme – F. de COULON/EPF, Lausanne

107 Généralités – signaux et systèmes – https://fr.slideshare.net/jbakkoury/chap1-generalitessignauxsystemes

108 Système – https://fr.wikipedia.org/wiki/ Système

109 Chapter 5: Digital Communication System - Digital Communication System BENG 2413 - Communication Principles Faculty of Electrical Engineering - http://slideplayer.com/slide/9933421/

tions and information between devices

- ✓ Examples: Modems, Interface Cards[110]
- ✓ Access point, smartphone

Functions of a communication device

- ✓ Transmission (transmitter)
- ✓ Reception (receiver)
- ✓ Conversion (converter)
- ✓ Adaptation (adapter)

Basic characteristics of a communication device

- ✓ Speed
- ✓ Coverage (range)
- ✓ Capacity

Basic steps of Telecommunications

4 basic steps

1.- Acquisition of information
2.- Processing of informations
3.- Transmission of information
4.- Use of information

1.- Acquisition or capture of user information

Techniques used to acquire the information to be transmitted (Information converted into a form suitable transmission systems)

- ✓ Use of interfaces to capture and transfer information to the transmitter
- ✓ Transformation of a physical quantity into an electrical signal (analogical chain) or a set of numerical values (digital chain)[111]
- ✓ Means of feeding the information transmission system
- ✓ Different interfaces for different information to be transmitted
    - o Microphone for sound and music
    - o Camera for images and videos
    - o Keyboard and screen for texts and data

110 Introduction to Communication Systems and Networks - https://web.sonoma.edu/users/f/farahman/sonoma/courses/es465/lectures/es465_fall2010/introduction.pdf

111 Tronc commun
Outils et méthodes d'analyse et de description des systèmes

2.- Processing of user information

- ✓ Set of actions to make signals (information envelopes) suitable for transmission and reception

Types of processing of information carried by a signal

- ✓ *Conversion*: Change of the nature of the signal for its processing and transmission
- ✓ *Digitization:* conversion of analog signals to the digital signal
- ✓ *Coding*: Transformation of information into coded information for processing
- ✓ *Decoding*: Operation of restoration of the signal to the state before the coding
- ✓ *Filtering*: Technique used to select the desired signals and frequency range
- ✓ *Amplification*: Increase of the amplitude of a signal
- ✓ *Modulation*: Transposition of the baseband signal spectrum into a higher frequency band to allow free space transmission
- ✓ *Demodulation*: Return of the signal in its base band for its use
- ✓ *Mixing* : Combination of several signal sources
- ✓ *Multiplexing*: Technique for combining several signals for simultaneous transmission on the same transmission medium
- ✓ *Demultiplexing*: Reverse multiplexing operation
- ✓ *Compression*: Reduction of the physical size of the information

Digitization of information

- ✓ Process of converting analog information (analog signals) into digital data (digital signals)
- ✓ Method for constructing a discrete representation of an object of the real world[112]
- ✓ Action to turn a document into a file readable by a digital device (computer, tablet, smartphone)

Digitization Examples

- ✓ Conversion of the human voice into a digital signal

112 http://www.techno-science.net/?onglet=glossaire&definition=321

- ✓ Transformation of paper document to a digital file by a scanner

Benefits of Digitization

- ✓ Easy storage, processing and restitution
- ✓ Multimedia integration
- ✓ Low error rate of digital links compared to analog links
- ✓ Cost of digital components (equipment) lower than that of analog components[113]
- ✓ Smaller components (miniaturized components)
- ✓ Better performing components
- ✓ Cheap components

Digital attributes

- ✓ Storage of digital information: optical discs (CDs, DVDs, BDs), hard disks, flash memory ...
- ✓ Digital transmission: digital circuits good steps, compact, possibility of digital processing of the received signal
- ✓ Increased immunity of digital signals to noise[114]

Rationale for digitization

- ✓ Digital Technologies: Telecom/ICTs base (bit-based systems operations)
- ✓ Better processing of all kinds of information (digitized information: information suitable for any processing)
- ✓ Better transmission of all kinds of information
- ✓ Better management of digitized information (consultation, indexing, storage, distribution, modification, updating, preservation)
- ✓ Availability of information anywhere (portable digital information)
- ✓ Cheaper

Conversion of analog signals to digital signals

Important step to facilitate the processing of analog information by digital systems

113 Liaison de données - Plan
https://www-npa.lip6.fr/~kt/ist/liaison

114 Télécommunications fixes et mobiles Module M1108
http://chamilo1.grenet.fr/ujf/courses/M1108CODAGEACQUISITIONETCODAGED/document/Transparents/slides-M1108.pdf

- ✓ *Goal : Use and processing of signals by digital systems*
- ✓ Transformation of analog information into digital data
  - o Conversion of human voice (naturally analog) into digital data

Conversion of digital signals to analog signals

Important step to facilitate the exploitation of digital information by human users

- ✓ *Goal: Signal transmission and use of information by the end user*
- ✓ Transformation of digital information into analog information
  - o Conversion of the digitized human voice into an analog signal

3.- Transmission of user information

- ✓ Set of devices for carrying information from one place to another

Implementation of remote communications (telecommunications)

- o *In wired mode: aluminum and copper cables, waveguide, optical fiber, etc.*
- o *In wireless mode: radiocommunication Wireless broadcasting, cellular link, microwave links using electromagnetic waves as media*

Information transmission chain

- ✓ Physical process of information transmission
- ✓ Set of devices for transporting information from one place to another[115]

Transmission types

- ✓ Baseband transmission or digital transmission
  - o Transmission of signals in their original frequency band
  - o Example: telephony: voice signals from the microphone (100 Hz ...... .5 KHz) and transmitted on symmetrical pairs
  - o Television: Video signals (50 Hz .... 5 MHz) from the camera and transmitted over coaxial cable

115 https://www.assistancescolaire.com/eleve/TS/physique-chimie/reviser-le-cours/

- o Data transmission: coded and formatted signals for transmission without frequency transposition[116]

Transposed band transmission or analog transmission

- o Transposed band transmission = transmission through modulation
- o Modulation = signal transposition from a lower band to a higher band of frequency
- o Impossibility to transmit in free space without modulation
- o Example: AM and FM audio signals: Transposition (from 100 Hz - 5 KHz) to 530 -1700 KHz (AM) and 88 -108 MHz (FM)

Information Transmission Process

- ✓ Acquisition
- ✓ Coding
- ✓ Transmission
- ✓ Decoding
- ✓ Restitution

Transmission media

2 environnements : physical and non physical

- ✓ Physical environment
  - o Guided propagation: signal propagation inside a metallic cable or waveguide, or optical fiber
  - o Example: cable television, wireline telephony, waveguide connecting a transmitter to a transmitting antenna
- ✓ Non -physical environment
  - o Unguided propagation: Propagation of the Electromagnetic Wave in Vacuum or Free Space
  - o Example: broadcasting, mobile telephony, satellite communication, mobile Internet

Transmission of telecommunications signals: 4 environments

- ✓ Subsea route: submarine cables carrying telecommunications signals
- ✓ Underground route: Telecommunications cables buried in the ground
- ✓ Aerial route : Cables on pylons carrying signals
- ✓ In free space: Signals propagating in free space without physical medium (radiocommunication)

Transmission by cable

- ✓ Wired transmission
- ✓ Cable transmission
- ✓ Transmission by waveguides
- ✓ Optical fiber transmission

Radio transmission

- ✓ Transmission by radio waves (Radiocommunication)

Transmission Media

- ✓ Physical path connecting a transmitter and a receiver in a transmission system
- ✓ Any means for transporting information in the form of signals from their source to their destination

Limits of transmission media

2 main limits

- ✓ Distance (coverage)
- ✓ Bit rate

Weaknesses of transmission media

3 main weaknesses

- ✓ Attenuation
- ✓ Phase shift
- ✓ Distortion

Transmission channel

- ✓ Division of a transmission medium dedicated to a link
- ✓ Physical medium used as a medium to transfer information between two remote points
- ✓ Route used as a vehicle for the transfer of information

Signal propagation

- ✓ Wired transmission (guided propagation)

Propagation of signals in a transmission cable

116 Systèmes de télécommunications, Bases de transmission P.-G. FONTOLLIET

- o Electric Cable (Coaxial cable, Twisted pairs)
- o Optical Fiber
- ✓ Radio propagation (radiated propagation)

Propagation of electromagnetic waves in free space

- o Transmission by radio waves
- o Micro wave links
- o Satellite transmission

Analog transmission

Analog signal transmission based on the combination of two fundamental elements:

1.- Modulating analog or digital signal or baseband signal or information signal

2.- Carrier signal (analog carrier or impulse carrier) generated by the transmitter for performing the "Modulation" operation

Advantages of analog transmission

- ✓ Simple system
- ✓ Cheaper

Disadvantages of analog transmission

- ✓ Noise in the signal
- ✓ Degradation of signal quality
- ✓ High bandwidth (compared to baseband transmission)[117]

Digital transmission

Transmission and processing of binary signals

- ✓ **Routing of a "digital" (or digitized) information through an "analog physical medium"**

2 options

1.- Transmission of binary signals in the binary state: baseband transmission (wired transmission)

2.- Modulation of the binary signal by an analog carrier: transmission in free space

Advantages of Digital Transmission

- ✓ Best quality
- ✓ Less distortion and attenuation
- ✓ Immunity to noise
- ✓ Multiplexing
- ✓ Larger coverage
- ✓ Low link error rates [118]

Disadvantages of digital transmission

- ✓ Greater bandwidth
- ✓ Coding and decoding circuits required
- ✓ Precise synchronization required

Types of transmission and information

4 possible scenarios

1.- Analog transmission of analog information

- ✓ Transmission of sound and images through radio waves
- ✓ Transmission of sound (telephony) through the PSTN (Public switched telephone network)

2.- Analog transmission of digital information

- ✓ Transmission of digitized data over telephone or satellite lines
- ✓ Data transport by modem and fax

3.- Digital transmission of digital information

- ✓ Local Area Networks, Integrated Services Digital Networks (ISDN)
- ✓ Transmission of digitized data on optical fibers, etc. (baseband transmission)

4.- Digital transmission of analog information

- ✓ Voice over IP (VoIP), Video conferencing over a local area network
- ✓ Transmission of speech, sound and images in baseband

Types of transmission medium

- ✓ Limited (palpable) medium: twisted pair, coaxial cable, fiber optic, waveguide
- ✓ Unlimited media: air (electromagnetic waves, infra-red)

117 Réseaux informatiques : télécoms & réseaux architectures et concepts fondamentaux - Christian Attiogbé http://pagesperso.lina.univ-nantes.fr/~attiogbe-c/mespages/RESEAUX/Licence/slides_general.pdf

118 Réseaux informatiques : télécoms & réseaux architectures et concepts fondamentaux - Christian Attiogbé http://pagesperso.lina.univ-nantes.fr/~attiogbe-c/mespages/RESEAUX/Licence/slides_general.pdf

Common characteristics of transmission media

- ✓ Bandwidth
- ✓ Noise and distortion
- ✓ Capacity
- ✓ Price
- ✓ Physico-chemical resistance to the environment
- ✓ Adaptation to the installation conditions

Wired transmission system

- ✓ Transmission provided by cables: twisted pairs, coaxial cables, optical fiber

Wired medium transmission chain

- ✓ Capture and conversion of the message into an electrical signal
- ✓ Electrical signal processing
- ✓ Transmission of the signal =
- ✓ Amplification of the signal
- ✓ Reception of the electrical signal
- ✓ Amplification
- ✓ Electrical signal processing (filtering, demodulation, etc.)
- ✓ Conversion of the electrical signal into an original message
- ✓ Restitution of the message to the recipient

Twisted pairs

- ✓ Cable consisting of two copper or aluminum wires wrapped around each other surrounded by plastic insulation
- ✓ Two intertwined copper strands
- ✓ Diameters: 0.4; 0.6; 08 or 1 mm

Constitution of twisted pairs

- ✓ 2 conductors
- ✓ 1 cable (envelope)
- ✓ 1 Insulation (protection against crosstalk)

Types of twisted pairs

2 types of twisted pairs

Type 1: Unshielded twisted pairs

- o Category 1: telephony
- o Category 2: voice and data transmission (4Mb/s)
- o Category 3: maximum bit rate of 10 Mb/s
- o Category 4: maximum bit rate of 16 Mb/s
- o Category 5 and extended category 5: maximum bitrate: 155 Mb/s
- o Category 6: maximum bit rate of 200 Mb/s
- o Category 7: maximum bit rate of 6 Gb/s

Type 2: Shielded twisted pairs

- ✓ Shielded cable consisting of a metallic braid
- ✓ Use of a better and more protective copper sheath[119]
- ✓ High speed and longer distances than UTP twisted pair

Signal and twisted pairs

- ✓ Medium designed and adapted to the transmission of electrical signals (currents or voltages)
  - o Electrical signals (signals obtained after conversion of information by transmission transducers)

Characteristics of twisted pairs

- ✓ Average attenuation (of the order of 0.1 dB/m)
- ✓ High characteristic impedance (between 150 ohms to 600 ohms)
- ✓ Paradiaphony (disturbance of one pair on another)[120]

Uses of twisted pairs

- ✓ Telephony (telephony networks)
- ✓ Data transmission (computer networks)

Coaxial cables

- ✓ Cable formed of two conductors centered on the same axis and separated by an insulating dielectric layer

---

119 Les techniques de transmission – Moula Malmoula http://www.academia.edu/7212786/Chapitre_3_LES_TECHNIQUES_DE_TRANSMISSION_La_couche_physique

120 Réseaux – Formation Telecom Réseaux – Pléneuf – V1.1 –Septembre 2011 – Edition numérique

- ✓ Asymmetrical transmission line used in high frequencies

Constitution of coaxial cable

- ✓ Core: element serving as channel of information transmission
- ✓ Cladding: outer environmental protection
- ✓ Shield: metal enclosure protecting data during transmission against noise
- ✓ Insulation: protective element avoiding contact with the shield)[121]

Types of coaxial cables

2 types of coaxial cables

Type 1: Thin coaxial cable (Thinnet or Cheapernet)

- ✓ Flexible wire
- ✓ Diameter: 6 mm
- ✓ Speed : 10 Mb/s
- ✓ Maximum length: 185 meters

Type 2: Thick coaxial cable (Thicknet ou Thick Ethernet)

- ✓ Rigid wire and shielded cable
- ✓ Diameter: 12 mm
- ✓ Speed: 10Mb/s
- ✓ Distance: 500 meters
- ✓ Use: Link for Ethernet networks, backbone for the connection between small networks

Signal and coaxial cables

- ✓ Medium designed and adapted to the transmission of electrical signals (currents or voltages)
  - o Electrical signals (signals obtained after conversion of information by transmission transducers)

Characteristics of coaxial cables

- ✓ Average attenuation (45 dB/Km at 10 MHz)
- ✓ Characteristic impedance: 50 ohms
- ✓ Maximum length of a segment (less than a few Km)[122]

Uses of coaxial cables

- ✓ Television
- ✓ Data transmission network
- ✓ Transmitter-transmit antenna link
- ✓ Connections between sound equipment
- ✓ Inter-urban telephone and submarine links

Optical Fiber

- ✓ Very thin glass or plastic cable having the property to conduct light
- ✓ Transmission medium used in telecommunications
- ✓ Light signal: Means of propagation in the optical fiber

Constitution of the optical fiber

3 Parts

- ✓ Core: Confinement and propagation of light energy (light signal, optical signal)
- ✓ Cladding: support to confinement of the signal within the core
- ✓ Protective coating: Mechanical protection of the fiber

Types of optical fiber

- ✓ Single mode fibers
- ✓ Multi - mode fibers (graded index or step index)

Signal and optical fiber

- ✓ Use of light signals or optical signals for the transmission of information
- ✓ Signals capable of traversing the optical fiber to carry information of all kinds
- ✓ Optical signals (signals obtained after conversion of electrical signals)

Characteristics of the optical fiber

- ✓ High bandwidth (broadband)
- ✓ Insensitivity to electrical and magnetic noise
- ✓ Small footprint
- ✓ Very low attenuation
- ✓ High propagation speed (in single mode)

---

121 Les techniques de transmission – Moula Malmoula http://www.academia.edu/7212786/Chapitre_3_les_techniques_de_transmission_La_couche_physique

122 Réseaux – Formation Telecom Réseaux – Pléneuf – V1.1 –Septembre 2011 – Edition numérique

- ✓ Security
- ✓ Lightness[123]
- ✓ reliability
- ✓ Increase of the distance between repeaters

Optical fiber transmission chain

- ✓ Capture and conversion of the message into an electrical signal
- ✓ Amplification
- ✓ Electrical signal processing
- ✓ Conversion of electrical signal into optical signal
- ✓ Optical signal transmission via optical fiber
- ✓ Possible amplification of the signal
- ✓ Conversion of the optical signal into an electrical signal at the reception
- ✓ Signal amplification
- ✓ Signal processing
- ✓ Conversion of the electrical signal into an original message
- ✓ Restitution of the message to the recipient

Uses of optical fiber

Use of light as a signal to convey information of all kinds

- ✓ Transport of all types of signals (voice, images, videos, etc.)
- ✓ Transmission network (backbone)
- ✓ Connection of subscribers
- ✓ Long-distance telecoms applications (single-mode fiber)
- ✓ Short-distance telecoms applications (Single mode fiber)

Conversion of the electrical signal into a light signal and vice versa

- ✓ Conversion of the emitted electrical signal into a light signal at the starting point of the optical fiber
- ✓ Conversion of the received light signal into an electrical signal upon arrival of the optical fiber at the ending point

Means used in the conversion

- ✓ Optoelectronic converter (emission): LED (Light Emitting Diodes) and LASER (Light *Amplification* of Stimulated Emission of Radiation)
- ✓ Optoelectronic converter (reception): Photodiode and phototransistor

Optical fiber transmission process

- ✓ Conversion of the electrical signal to be transmitted into a light signal,
- ✓ Transmission the light signal via optical fiber
- ✓ Conversion of the light signal into an electrical signal at the other end of the link and
- ✓ Processing of the electrical signal for proper use

Deployment of Optical Fiber

3 deployment modes

- ✓ *Aerial deployment*: Installation of optical fiber on pylons throughout the area to be served
- ✓ *Underground deployment*: Installation of optical fiber in the soil
- ✓ *Underwater deployment*: Installation of optical fiber in the seabed (in the ocean)

Submarine cables

- ✓ Cables laid on the seabed, intended to carry telecommunications signals or to carry electrical energy
- ✓ Cables used to transport telecommunications signals (telephony, Internet) between two countries or two continents
- ✓ Alternative solution to satellite links connecting very remote points

Types of submarine cables

- ✓ Electrical cables: coaxial cables and twisted pairs
- ✓ Optical cable: optical fiber

Laying of submarine cables

- ✓ Operation carried out using a suitable ship, called a deep -sea vessel
- ✓ *Deep -sea vessel*: a ship specialized in the installation, lifting and maintenance of submarine telecommunications cables

123 Réseaux – Formation Telecom Réseaux – Pléneuf – V1.1 –Septembre 2011 – Edition numérique

- ✓ *Station of departure*: Starting point of cable deployment
- ✓ *Landing Station (Terminal Station)*: Point of arrival of the submarine cable on the ground

Wireless transmission systems

Transmission carried out thanks to the propagation of radio waves in the free space.

- ✓ Radiocommunication : Remote communication using radio waves (electromagnetic waves)
- ✓ Radiocommunication: communication achieved without physical medium

Wireless transmission medium

Point-to-point link and Broadcasting by radio waves (radiocommuniation)

- ✓ Transmission and Broadcasting of signals through the air over long distances

Principles of radiocommunication

- ✓ Use of free space as a medium for the propagation of radio waves
- ✓ Use of transmit and receive antenna for broadcasting

Characteristics of wireless transmissions

- ✓ Average bandwidth
- ✓ Sensitivity to electrical and magnetic noise
- ✓ Very small congestion and weight
- ✓ Performance highly dependent on transmission conditions
- ✓ Security and reliability difficult to ensure[124]

Radioelectricity

- ✓ Study of radio transmission, propagation of waves, and interfaces with transmitter and receiver via antennas[125]
- ✓ Study of the production, the emission and the transmission of the radio waves
- ✓ Basis for all electronic communications techniques performed without physical medium

Applications of radiocommunication

- ✓ Radiotelephony (wireless telephony)
- ✓ Broadcasting
- ✓ Microwave links
- ✓ Satellite links
- ✓ Radar
- ✓ Amateur Radio
- ✓ Etc.

Wireless transmission chain

- ✓ Capture and conversion of the message into an electrical signal
- ✓ Amplification and electrical signal processing
- ✓ Transfer via cable or connector of the electrical signal to the transmit antenna
- ✓ Transformation of the electrical signal into electromagnetic waves
- ✓ Propagation of electromagnetic waves in free space
- ✓ Possible use of passive or active repeater
- ✓ Reception of electromagnetic waves by a receive antenna
- ✓ Transformation of electromagnetic waves into electrical signal
- ✓ Transfer of the electrical signal to the receiver via a cable
- ✓ Amplification of the electrical signal
- ✓ Electrical signal processing
- ✓ Conversion of the signal of the electrical signal into an original message
- ✓ Restitution of the message to the recipient

Telecommunications antennas

- ✓ Fundamental element in radiocommunication systems
- ✓ Device capable of transmitting and receiving electromagnetic waves (radio waves or radio waves)

---

124 Réseaux – Formation Telecom Réseaux – Pléneuf – V1.1 –Septembre 2011 – Edition numérique

125 Gestion d'interconnexion et dérégulation de flux d'appel dans un serveur téléphonique elastix https://www.memoireonline.com/09/14/8917/Gestion-dinter-connexion-et-deregulation-de-flux-dappel-dans-serveur-telephonique-elastix.html

- ✓ Interface between a guided propagation medium (coaxial or two-wire line) and an unguided propagation medium (free space)
- ✓ Interface between the transmission system and the free space (the air)
- ✓ Interface between reception system and free space (air)

Functions of transmitting and receiving antennas

Transmission and reception functions

Transmission

- ✓ Conversion of electrical energy into electromagnetic waves
- ✓ Radiation of electromagnetic waves in free space

Reception

- ✓ Capture of radiated electromagnetic waves
- ✓ Conversion of electromagnetic waves into electrical signal through induction

Principle of antennas

- ✓ Trasmit side: Radiation of electromagnetic energy in the free space by any conducting wire or any metal traversed by a variable current
- ✓ Receive side: Generation of an induced current by any conductor or metal hit by electromagnetic waves

Types of antennas[126]

Antennas	Types	Use
Directional antennas	Parabolic antena Yagi antennas	Microwave links Satellite communication
Omnidirectional antennas	Vertical strand	Radio, Television Mobile communication

Space Telecommunications

- ✓ Telecommunications provided by a system installed in space
- ✓ Telecommunications with infrastructure installed in space (satellite)
- ✓ Means of communication between earth and space
- ✓ Example: Telecommunications Satellite

Telecommunications Satellite

- ✓ Radio relay in orbit for the transmission of information of all kinds between two remote points on the earth via earth stations
- ✓ Means of transmitting and receiving information by electromagnetic waves

Links of Telecommunications satellite

- ✓ Uplinks: Radio links connecting the transmitting earth station to the satellite
- ✓ Downlinks: Radio links connecting the satellite to the receiving earth station
- ✓ Inter-satellite links: Links established between two telecommunications satellites

Justification of a satellite as a means of transmission

- ✓ Largest radio coverage thanks to its altitude
- ✓ Coverage of remote areas
- ✓ Fewer repeaters installed on the ground
- ✓ Less risk to infrastructure during natural disasters (Main element (satellite) in orbit)
- ✓ Means to avoid obstacles (mountains, trees, buildings)
- ✓ High throughput

Components of a telecommunications satellite

Payload and platform

1.-**Payload**

- ✓ Component acting as an active radio relay
- ✓ Composed of transmitters, receivers, amplifiers, antennas and other accessories

2.-**Platform (load module)**

Infrastructure with the following functions

126 Antennes
https://fr.scribd.com/doc/21285105/Cour-d-antennes

- ✓ Supply of electrical energy (solar panels)
- ✓ Thermal regulation (thermal control)
- ✓ Positioning and control of the satellite placed in space (Propulsion systems for remote control maneuvers, stabilization, orbit control)

Uses of telecommunications satellites

- ✓ Fixed and mobile telephony
- ✓ Radio and Television
- ✓ Data transmission, Internet
- ✓ Transmission network (backbone) for networks
- ✓ Geolocation

Stakeholders in satellite telecommunications

- ✓ *Manufacturer*: satellite manufacturing company (Alcatel, Hugues, Lockheed Martin, Loral Space Systems)
- ✓ *Launcher*: company specialized in the launch of artificial satellites (Ariane Space, Delta, Titan, Atlas Centaur)
- ✓ *Satellite Operators*: Enterprise managing satellites to provide telecommunications links (Intelsat, Eutelsat, Arabsat)
- ✓ *Satellite services Operators*: Operators of telephony or satellite television (Iridium, Inmarsat, Globaltar, CanalSat, Canal +)
- ✓ *Telecommunications Operators*: Operators leasing satellite links for the provision of services (telephony, television, satellite, (international telephone calls, Internet)
- ✓ *Eartn Station Operator*: operator specialized in the establishment of connections with telecommunications satellites
- ✓ Consumers of telecommunications services

Frequencies used by communications satellite

- ✓ 6/4 GHz bands (Band C)
- ✓ 6 GHz band for uplinks
- ✓ 4 GHz band for downlinks
- ✓ 14/10-12 GHz bands (Ku Band)
- ✓ 30/20 GHz bands (Ka Band)

Types of telecommunications satellites

Fixed satellite and orbiting satellite

1.- Geostationary or geosynchronous satellites

- ✓ Altitude: 36,000 km (35,863 km more precisely)
- ✓ Coverage: 1/3 of the earth (3 satellites Geo: cover of the whole planet earth)
- ✓ Uses
  - o Weather forecast
  - o Communications (Broadcasting, Global Communications, Military Communications

2.- MEO: Medium Earth Orbit (Medium altitude satellite)

- ✓ Altitude: 10,000 - 15,000 km
- ✓ Approximately 15 MEO for overall coverage
- ✓ Uses
  - o Navigation (GPS, Galileo, Glonass)
  - o Communications (Inmarsat)

3. - LEO: Low Earth Orbit (Low Altitude Satellite)

- ✓ Altitude: 700 - 2000 Km
- ✓ More than 30 LEO for global coverage
- ✓ Uses
  - o Earth observation (Google earth)
  - o Communications (Iridium, Globalstar)
  - o Search and rescue

4.- HEO: Highly Elliptical Orbit

- ✓ Altitude: 38000 km - 50000 Km
- ✓ Great coverage
- ✓ Uses
  - o Satellite Radio
  - o Data recognition

Passive satellites

- ✓ Wireless relay only acting as a reflector for signals received from the earth station (first satellite in orbit)

Functions of passive satellite

- o Reception of the signal from an earth station

- o Reflection of the signal to the receiving earth station

Active satellites

- ✓ Radio relay with amplification, processing and re-transmission capability of signals received from the earth station

Functions of active satellite

- o Reception of a signal from an earth station
- o Amplification of the received signal
- o Possible change of frequency of the signal
- o Re-transmission of the processed signal to another earth station

Orbit and frequency allocation for satellite communications

- ✓ Orbit allocated by ITU (International Telecommunication Union)
- ✓ Uplink and downlink frequencies assigned by the ITU

Earth stations

- ✓ Interface between the terrestrial networks and the telecommunications satellite installed in space
- ✓ VSAT (Very Small Aperture Terminal): Micro earth station installed at users' prmises for access to satellite television and Internet signals

Roles of an earth station

- ✓ Transmission of signals received from telecommunications networks to telecommunications satellites
- ✓ Reception of signals emitted by telecommunications satellites and routing to telecommunications networks
- ✓ Reception of signals from television stations of other countries facilitated by satellite reception antennas, commonly called dish

Components of an earth station

- ✓ Radio transmitter/receiver
- ✓ Antennas and feeding cables
- ✓ Signal processing equipment
- ✓ Low Noise Block - Converter (LNB)

Infrared links

- ✓ Infrared: electromagnetic wave of higher frequency than light
- ✓ Communication established using infrared rays
- ✓ Wireless transmission medium carrying information via light beams

Principle of infrared links

- ✓ *Emission*: Electroluminescent diodes (infrared radiation emission) / conversion of the electrical signal into infrared radiation by an LED)
- ✓ Modulation of infrared beams generated by LED
- ✓ Sending of signals through the air via light waves
  - o Line of sight Transmission (transmitter and receiver in visibility over short distances)
- ✓ *Reception*: use of silicon photodiodes to convert infrared radiation into electrical signal

Characteristics of infrared links

- ✓ Speed: 10 Mb/s
- ✓ Distance: 30 meters

Uses of infrared links

- ✓ Remote controls
- ✓ Connection between multiple devices (laptops)
- ✓ Wireless LANs
- ✓ Wireless modems
- ✓ Connection between computers and peripherals (wireless keyboards, wireless mouse, etc.)
- ✓ Short-distance data links between computers or mobile phones
- ✓ Short-distance data transmission in the field of robotics
- ✓ Monitoring and control applications
- ✓ Missile guidance systems
- ✓ Security lights

Comparison between wired transmission and wireless transmission[127]

Advantages and disadvantages of wired communication

Advantages	Disadvantages
Reliability (not affected by other wireless signals: cellular telephones, microwave links)	Lack of protection against mold and other climatic conditions
Lower cost	Lack of protection against noise from machinery and magnetic fields
Long life expectancy	Limited length of cables
Broadband	
Quality of service	

Advantages and disadvantages of wireless communications (wireless transmission)

Advantages	Disadvantages
Convenience	Lack of reliability (Affected by other wireless signals: cellular telephones, microwave links)
Range	Interception of signals
Long life expectancy	Limited speed in wireless transmission
	Low quality of service

Comparison of different transmission media

Medium	Twisted pairs	Coaxial cables	Radio waves	Infra-Red	Optical Fiber
Propagation	guided	Guided	unguided	unguided	guided
Material	copper	copper	-	-	Silica, polymers
Bandwidth	KHz - MHz	MHz	GHz	GHz	THz
Attenuation	strong	strong with frequency	variable	obstacles	very weak
Sensitivity, electromagnetic disturbances	strong	Weak	strong	strong	null
confidentiality	limited	correct	null	relative	high
Energy transport	yes	yes	no	no	experimental
Interface cost	very weak	weak	rather weak	average	high
Support cost	very weak	high	null	null	high
Applications	telephone, low and medium speeds, networks, high speed over short distances	high speed local networks, video	mobile, satellites, wireless	remote control, "indoor" communications	high speeds long distances

127 Wired and wireless technologies
https://fr.slideshare.net/AKHILSabu1/wired-and-wireless-technologies

Bit rate of wired transmission media

- ✓ Twisted pair: 16 Mb/s
- ✓ Coaxial cable: 10 Mb/s
- ✓ Optical fiber: up to Tb/s

Bit rate of wireless transmission media

- ✓ Broadcasting: up to 2 Mb/s
- ✓ Microwave links: 45 Mb/s
- ✓ Satellite links: 50 Mb/s
- ✓ Cellular system: 100 Mb/s
- ✓ Infrared system: up to 4 Mb/s

Time and telecommunications

- ✓ *Time*: sensitive factor in signals transmission
- ✓ *Time:* very sensitive parameter in real-time communications
- ✓ A second: 7,5 turns of the earth by light (300 000 km/sec)
- ✓ Multiple and long operations in one second (millisecond and microsecond)
- ✓ Speed of the signal in metallic and optical cables lower than 300 000 km per second
- ✓ *Propagation time*: Time required for the signal to travel from one point to another

Propagation time and delay and delay introduced by telecommunications equipment

Propagation time: time of transfer of information from source to destination

- ✓ Optical fiber: 0.54 millisecond per 100 km
- ✓ Microwave transmission: 0.38 millisecond per 100 km
- ✓ Coaxial cable: 0.42 millisecond per 100 km
- ✓ Copper cable: 1.08 millisecond per 100 km
- ✓ Geostationary satellite: 240 milliseconds (2 by 120 millisecond)
- ✓ Low altitude satellite: 10 milliseconds (2 by 5 millisecond)

Delay due to Equipment (in one direction only)

- ✓ Analog multiplexing: 2 milliseconds
- ✓ ATM switch: 1.6 millisecond
- ✓ Digital switch: 0.5 to 1.2 milliseconds
- ✓ Digital compression: 5 to 150 milliseconds
- ✓ MPEG -2 encoder: 150 milliseconds
- ✓ MIC equipment: 0.125 to 0.5 milliseconds
- ✓ Digital connection: 1.2 millisecond
- ✓ Mobile phone : 10 millisecondes (DECT), 90 millisecondes (GSM)[128]

Distance and telecommunications

Attenuation of the signal directly proportional to the distance traveled

- ✓ *Distance:* source of attenuation of the signal being transmitted
- ✓ *Distance*: more investment in telecommunications infrastructure
- ✓ *Distance*: more resources (human, technical, energy, logistic) for the provision of the service
- ✓ *Distance*: Technical implications (amplifiers and active relays to be installed on the links)

Criteria of choice of a transmission medium a transmission medium

- ✓ Cost (acquisition and maintenance)
- ✓ Transmission capacity (number of channels available)
- ✓ Transmission quality (reliable reproduction of information without interference, etc.)
- ✓ Distance (range)
- ✓ Bitrate (Bandwidth)
- ✓ Attenuation (level of attenuation of the signal)
- ✓ Immunity to noise and interference
- ✓ Easy deployment
- ✓ Lifetime

4. Use of user information

Use of the conveyed information

Means

128 Cours B11 – Transmission des Télécommunications – Partie 2 – Chapitre 2

- ✓ Reception of signals by the receiver
  - o Adjustment of the receivers to the signal transmission frequency
- ✓ Processing of electrical signals by the receiver
- ✓ Use of interfaces for the use of information
  - o Speaker for sound and music
  - o Screen for images and videos
  - o Suitable screen for texts and data

TELECOMMUNICATIONS TECHNOLOGIES

Network, transmission and communication technologies

Network Technologies

- ✓ Wi-fi
- ✓ Token Ring
- ✓ Frame Relay
- ✓ PSN (Packet Switching Network)
- ✓ ISDN (Digital Integrated Services Network)
- ✓ ATM (Asynchronous Transfer Mode)
- ✓ FDDI (Fiber Distributed Data Interface)
- ✓ SONET (Synchronous Optical Network)
- ✓ DDN (Digital Data Network)
- ✓ Ethernet
- ✓ Etc.

Digital Network Technologies

- ✓ Digital transmission technologies transmitting information under discrete impulse

Benefits of digital network technologies

- ✓ Largest transmission bit rates
- ✓ Transmission of large amounts of information
- ✓ Biggest economy
- ✓ Low error rate

Wired transmission technologies

- ✓ Powerline Communication
- ✓ Optical Fiber
- ✓ Cable Modem
- ✓ Digital Subscriber Line (DSL)

Wireless transmission technologies

- ✓ Radio broadcasting
- ✓ Microwave links
- ✓ Satellite
- ✓ Bluetooth
- ✓ Cellular technologies (2G, 2.5G, 3G, 4G, 5G)
- ✓ Wi-fi (Wireless Fidelity)
- ✓ Wimax (Worldwide Interoperability for Microwave Access)
- ✓ FSO (Free Space Optics)
- ✓ LMDS (Local Multipoint Distribution Service)
- ✓ Multi Channel Multipoint Distribution Service (MMDS)
- ✓ Infra red
- ✓ Wireless Power Transfer

Switching technologies

- ✓ Circuit switching (telephony)
- ✓ Packet switching (Internet)
- ✓ Message switching
- ✓ Frame relay switching
- ✓ Cell switching

Applications of telecommunications technologies

Use of these technologies in the provision of the following services

- ✓ Telephony
- ✓ Video conference
- ✓ Email
- ✓ Instant messaging
- ✓ Voice Messaging

- ✓ Chat room (chat rooms, chat rooms)
- ✓ Newsgroup
- ✓ Collaboration
- ✓ Group work software (groupware, groupware)
- ✓ Geolocation system (GPS)

Data network technologies

- ✓ Token Ring
- ✓ 802.11
- ✓ 802.11n
- ✓ WiMAX
- ✓ Bluetooth
- ✓ Zigbee

TELECOMMUNICATIONS NETWORKS

- ✓ Electronic systems consisting of links and switches, including controls of operations, information transfer and exchange between multiple users
- ✓ Means of communication between a transmitter and a receiver
- ✓ System making possible communications from the transmitter to the receiver
- ✓ Set of transmitters, receivers and communication channels to exchange information
- ✓ Means of communication between remote or multiple users (man / machine) of a service (telephony / fax / Internet) exchanging information (voice, data, images, ...) via a terminal (touch-tone telephone, fax)[129]
- ✓ Combination of telecommunications and computing resources to facilitate remote communication
- ✓ Combination of telecommunications and computing resources for information sharing between distant points
- ✓ Network of arcs (telecommunication links) and nodes (switches, routers ....) set up for the transmission of messages from one end to the other through multiple links[130]
- ✓ System designed for transmitting signals of all kinds using electrical and electromagnetic energy

Rationale for a telecommunications network

- ✓ Connection of users
- ✓ Broadcasting of information
- ✓ Access to remote resources (servers, printers)

Types of telecommunications network

- ✓ Broadcasting Network: Transmission from a Source to Many Users
  - o Sound and television broadcasting
  - o Examples: radio and television stations
  - o Means: Established Permanent Links and Terminals
- ✓ Collection network: Transmission from multiple sources to the same recipient
  - o Telemetry (telemetry)
  - o Example: Ocean Monitoring Network
  - o Means: Established Permanent Links and Data Collector
- ✓ Switched network: links established on the order of the user
  - o Telephony
  - o Example: Telephone network
  - o Means: wired and wireless transmission systems, switches and terminals

Fundamental elements of a telecommunications network

2 basic elements

- ✓ Nodes
  - o Routing of information
  - o Switching of information
  - o Network management

129 Réseau http://www.etudier.com/dissertations/Reseau/518363.html

130 Gestion de réseaux de télécommunications http://helpforafricanstudents.org/francais/index.php/message-introductif-important/42-carrieresprofessionelles-/communications-c/122-gestion-de-reseau-des-telecommunications-

- ✓ Links
  - o Interconnection of the nodes
  - o Transport of information[131]

Examples of telecommunications networks

- ✓ Television network
- ✓ Switched telephone network
- ✓ Computer network
- ✓ Internet network

Main components of a telecommunications network

- ✓ User Terminals for access and use of the Service
- ✓ Computers for processing information and interconnected by the network
- ✓ Processor or processing machine (devices performing controls and supporting functions: switch, router, hub, gateway)
- ✓ Channels of transmission (channel of transmission of information)
- ✓ Telecommunications links forming a communication channels between a transmitting terminal and a receiving terminal
- ✓ Telecommunication equipment designed to facilitate the transmission of information
- ✓ Telecommunication software for controlling transmission of messages across the network
- ✓ Interfaces between equipment and users
- ✓ Hubs
- ✓ Multiplexers

Telecommunications network and the user of services

2 types

- ✓ Network Services
  - o Ability to support telecommunications services: telephony, video conferencing, e-mail, etc.
  - o Ability to respond to service requirements (bandwidth, real-time constraints)
  - o Alternatives (circuit switching, packet switching)[132]
- ✓ Network management
  - o Perpetual evolution of the network
  - o New subscribers to connect
  - o New materials to install
  - o New services to introduce
- ✓ Network operation
  - o Maintenance operations
  - o Traffic observation operations
  - o Quality of service observation operations[133]

131 Trafic et performances des réseaux de télécoms - Georges Fiche et Gérard Hébuterne http://197.14.51.10:81/pmb/TELECOMMUNICATION/Trafic%20et%20performances%20des%20reseaux%20de%20telecoms.pdf

132 Trafic et performances des réseaux de télécoms - Georges Fiche et Gérard Hébuterne http://197.14.51.10:81/pmb/TELECOMMUNICATION/Trafic%20et%20performances%20des%20reseaux%20de%20telecoms.pdf

133 Trafic et performances des réseaux de télécoms - Georges Fiche et Gérard Hébuterne http://197.14.51.10:81/pmb/TELECOMMUNICATION/Trafic%20et%20performances%20des%20reseaux%20de%20telecoms.pdf

Information, telecoms networks and services[134]

Type of information	Network	Service
Speech	Broadcasting network, switched network, ordinary network	Telephony (Conference call, Speaking clock) Broadcasting Intercom
Music	Broadcasting network	Broadcasting
Texts	Point-to-point fixed network, switched network, ordinary network	Telegraphy, telex, teletex, e-mail
Still images	Broadcasting network, point-to-point fixed network, switched network, ordinary network	Facsimile, videotext
Moving images	Broadcasting network, fixed point-to-point network, switched and ordinary network	Television, videophone
Data	Collection network, point-to-point fixed network, switched network, ordinary network	Teleinformatics (telemetry, remote monitoring, remote control)

134 Systèmes de télécommunications, Bases de transmission – P.-G. FONTOLLIET

Relations between the network and the terminals

- ✓ Physical or intangible link connecting the network to the terminal
- ✓ Adjustment on the same frequency
- ✓ Compatibility between the network and the terminals

Functions of a telecommunications network

- ✓ Connection of users
- ✓ Connection of users with servers
- ✓ Provision of telecommunications services
- ✓ Management of telecommunications services[135]

Fundamental components of a telecommunications network

- ✓ *Core Network*: Infrastructure enabling the interconnection of users with each other
- ✓ *Access network*: Link between the user and the core network
- ✓ *User Terminal (User Terminal Equipment)*: Means of access to services

Means implemented in telecommunications networks

- ✓ Organs for translation of the message into an electrical signal, and vice versa (electrical transducers)
- ✓ Signal Conditioning Electrical Circuits
  - o Compatibility of the signal with the transmission channel
  - o Simultaneous routing by multiplexing of several messages on the same channel (amplification, modulation, demodulation, filtering, etc.)
- ✓ Communication path selection equipment (central switch)
- ✓ Transmission channels (lines, radio links, etc.)[136]

Availability of telecommunications networks

- ✓ Permanently available systems (radio station, TV, telephony, Internet)

135 Trafic et performances des réseaux de télécoms - Georges Fiche et Gérard Hébuterne http://197.14.51.10:81/pmb/TELECOMMUNICATION/Trafic%20et%20performances%20des%20reseaux%20de%20telecoms.pdf

136 Cours/Lecture series – 1986 -1987 Academic Training Programme – F. de COULON/EPF, Lausanne

- ✓ Connection and use of services as needed (adjustment to the frequency of a radio station, dialing of a telephone number, sending of a text message, launching of a web browser for access to the Internet

Phases of a telecommunications network
- ✓ Network Design
- ✓ Network planning
- ✓ Network implementation (deployment)
- ✓ Network optimization

Mission of a network operator
- ✓ System design
- ✓ Installation
- ✓ Setting up
- ✓ Use
- ✓ Maintenance

Benefits of digitization of the Telecommunications Network
- ✓ Use of simpler, more economical and more integrated electronic components compared to analog components previously used
- ✓ Time multiplexing simpler and more reliable than frequency multiplexing used on long distance analogue lines
- ✓ Better adaptation to different information to convey (Sounds, images, data)
- ✓ Signaling easier to convey by incorporation in the data
- ✓ Easier regeneration of signals and no additional breath or noise (much lower error rate than analog)
- ✓ Signals easier to process

Disadvantages of the digitization of the telecommunications network
- ✓ Greater media bandwidth.[137]

Stakes related to the management of a telecommunications network
- ✓ Capacity planning and demand forecasting
- ✓ Network update
- ✓ Data management problem
- ✓ Network monitoring
- ✓ Optimum use of resources
- ✓ Management of customer problems
- ✓ Problem in decision-making

Value Added Networks (VAN)
- ✓ Closed network providing a set of users with telecommunications services of a quality different from that provided on the public network (most often higher), and allowing access to proprietary applications[138]
- ✓ Network adding value by processing the information conveyed

Examples of value-added networks
- ✓ SITA (International Aeronautical Telecommunication Society)
- ✓ SWIFT (Society for International Interbank Financial Telecommunications)

Telecommunications Network Management Model
2 management models

1.- TMN (Telecommunications Network Management)

2. - SNMIP (Simple Network Management Protocol)

Features of a network management application
- ✓ Supervision of network operation
- ✓ Signaling of breakdowns
- ✓ Detection of errors and anomalies
- ✓ Cost evaluation
- ✓ Recovery of statistical data[139]

Network Administration Functions
5 functions to meet functional needs

1- Management of anomalies
- ✓ Network monitoring

137 Les signaux http://aldevar.free.fr/data/14-Telephonie/signaux.pdf

138 Introduction aux télécommunications http://www.volle.com/ENSPTT/introtcom.htm#réseau à valeur ajoutée

139 TMN: Telecommunication management network https://wapiti.telecom-lille.fr/commun/ens/peda/options/st/rio/pub/exposes/exposesrio1998ttv/TMN/EXPOTMN1.HTM#Principes

- ✓ Anomaly detection
- ✓ Location of breakdowns
- ✓ Tests and measurements

2- Configuration management (management of configuration parameters)

- ✓ Component installation function
- ✓ Control and monitoring function
- ✓ Name Management Function
- ✓ Commissioning management
- ✓ State Management
- ✓ Order management

3- Security Management

- ✓ Protection
- ✓ Authentication management
- ✓ Management of the level of authorization

4- Performance Management

- ✓ Collection of data
- ✓ Traffic Management
- ✓ Quality of service management

5- Accounting information management

- ✓ Collection of statements of account
- ✓ Management of billing settings[140]
- ✓ Declaration of subscribers and terminals
- ✓ Billing
- ✓ Statistics

Management of a telecommunications network

- ✓ Traffic Management
- ✓ Security
- ✓ Network monitoring
- ✓ Capacity planning

140 Tmn: Telecommunication management network https://wapiti.telecom-lille.fr/commun/ens/peda/options/st/rio/pub/exposes/exposesrio1998ttv/TMN/EXPOTMN1.HTM#Principes

Layers of a network management application

5 layers

1- Business Management Layer

- ✓ Business aspects
- ✓ Marketing aspects and market shares
- ✓ Legislative aspects
- ✓ Return on investment
- ✓ Customer satisfaction
- ✓ Achievement of community and government goals

2- Service Management Layer

- ✓ Interface between service users
- ✓ Management of services provided to customers
- ✓ Quality of service
- ✓ Costs and duration related to market objectives

3- Network Management Layer

- ✓ Network aspects of management
- ✓ Routing and addressing
- ✓ Addition, deletion, and change capabilities

4- Element Management Layer

- ✓ Management of elements including networks and systems
- ✓ Management of the load of the elements by the planning of
    - o Sequencing, contradictory demands
    - o Contracting demands

5- Network Element Layer

- ✓ Switches management aspects
- ✓ Aspects related to the management of transmission media
- ✓ Aspects related to the management of distribution systems

Telecommunications equipment

- ✓ Physical element for providing and accessing telecommunications services (network equipment and terminals)

- ✓ Tool for overcoming time and distance
  - o Examples: Transmitter, receiver, antennas, router, switch, multiplexer, tablet, phone, digital tablet

Transmission equipment

- ✓ Transmission lines
- ✓ Optical Fiber
- ✓ Radio transmitter
- ✓ Telecommunications Satellite
- ✓ Etc.

Terminal equipment (User equipment)

- ✓ Cellular telephones
- ✓ Subscriber handset
- ✓ Radio and television receiver
- ✓ Computer
- ✓ Modem
- ✓ Fax machine
- ✓ Private Branch Office
- ✓ Router
- ✓ Etc.

Integrated Services Digital Network (ISDN)

- ✓ ISDN: Network providing end-to-end digital connectivity with a wide variety of services[141]
- ✓ ISDN: Network designed to associate voice, data, video and any other application or service
- ✓ ISDN: Telecommunications network consisting of digital links
- ✓ Integration of services = use of part or all of the telecommunication network for the *joint* transmission of various information relating to services of a different natures
- ✓ ISDN: Network entirely constituted by digital connections, and allowing its users to exchange information of different natures : sounds, images, data
- ✓ ISDN: Direct consequence of the digitization of the transmission and switching network
- ✓ ISDN: Transmission by subscriber of different types of information via a single interface
- ✓ Services available: Caller ID, dual calling, transmission of mini-messages, etc.)
- ✓ Available bitrate: 64Kbps at 2Mbps

Examples of integrated services

- ✓ Telephony
- ✓ Teleinformatics
- ✓ Telex and Teletex
- ✓ Fax
- ✓ Video Calling
- ✓ Broadcasting of musical or television programs
- ✓ Teleactection, telemetry, remote alarm, etc.[142]

141 Technologie RNIS – Philippe Latu – https://www.inetdoc.net/pdf/rnis.pdf

142 Systèmes de télécommunications, Bases de transmission –P.-G. FONTOLLIET

CHAPTER 5

TRADITIONAL AND IP TELEPHONY
CELLULAR TELEPHONY

TELEPHONY

- ✓ Transmission of the human voice and sound between two distant locations
- ✓ Telecommunication system essentially providing transmission and reproduction of speech (and more rarely other sound signals)

Types of telephony

Fixed telephony and mobile telephony

Fixed telephony (traditional telephony)

- ✓ Use of fixed networks
- ✓ Provision of basic fixed services

Mobile telephony

- ✓ Use of radiocommunication networks for the provision of telephony service (radiotelephony and paging)
- ✓ Provision of telephony services to mobile subscribers

TDM telephony and IP telephony

TDM (Time Division Multiplexing)

- ✓ Circuit switching (allocation of a circuit for the duration of the communication)
- ✓ Time multiplexing

IP (Internet Protocol)

- ✓ Packet switching (Transformation of digitized messages into packets and forwarding packets from node to node)
- ✓ Statistical multiplexing

General information on Telephony

- ✓ *Subscriber side*: Fixed or cellular terminal
- ✓ *Connections*: wired and wireless links
- ✓ Unique telephone number to each subscriber
- ✓ Centralization of the intelligence of the telephone system

Principles of telephony

- ✓ Conversion of sound messages into electrical signals by a microphone (transmission)
- ✓ Use of circuit switching for voice services
- ✓ *Circuit*: physical path of communication between two or more points of the network
- ✓ Operation in connected mode (establishment of a circuit between the two correspondents)
- ✓ Routing via a medium of the electrical signals to the called subscriber
- ✓ Conversion of electrical signals into sound messages by loudspeaker (reception)

Conditions of access to the telephone service

Access conditioned by

- ✓ Availability of a telephone terminal
- ✓ Connection to the telephone network
- ✓ Power supply of the telephone
- ✓ Availability of credit on the account or monthly subscription
- ✓ Contact details (telephone number)

Modes of access to fixed telephony

3 access modes

- ✓ Traditional telephone line (PSTN)
- ✓ ADSL connection via the Internet (Software installed on a computer via Internet / ISP)
- ✓ Telephony via cable (A connection of a wired network via a modem -cable provided by Cable Operators)[143]

Connection of subscriber to the telephone network

Local loop or last mile

- ✓ Local loop: Wired or wireless network segment linking terminal equipment to switching equipment
- ✓ Local loop: Wired or wireless link established between the customer premise equipment and the switch
- ✓ Means of connection
  - o Coaxial cables and twisted pairs
  - o Optical Fiber
  - o Cellular link
  - o Wimax
  - o Satellite (mobility and difficult access areas)

143 Guide pratique des communications électroniques http://www.economie.gouv.fr/files/files/conseilnationalconsommation/guide_interactif_securise.pdf

User Terminal equipment or Customer premise equipment

- ✓ Access and use of telecommunications services
- ✓ Example : Telephone, digital tablet, computer, etc.

Parts of a telephone

Microphone: Conversion of sounds into electrical signals

- ✓ Loud speaker: Conversion of the electrical signal into sound
- ✓ Pad: Telephone Number Dialing
- ✓ Hook - switch: Sending of alert to the central exchange for each connection request
- ✓ Alarm: Subscriber alert for incoming call
- ✓ Balancer circuit: Circuit used for echo prevention

Subscriber Link

- ✓ Line between the public network (home switch) and the subscriber installation

Departure Lines

- ✓ Lines carrying only outgoing calls
  - o Telemarketing lines

Arrival Lines (Specialized Arrival Point B)

- ✓ Lines routing only incoming calls
  - o Emergency lines
  - o Customer Services

Two-way telephone lines

- ✓ Telephones Lines carrying indifferently incoming or outgoing calls[144]

Telephone dialing in fixed mode

2 types of telephone keypad

- ✓ Pulse dialing
- ✓ Touch -tone dialing (DTMF: Dua-tone multi-frequency

Telephone services

2 categories: Basic services and value-added service

Basic telephone services

- ✓ Telephone calls
- ✓ Conference call

Value-Added Telephone Services

- ✓ Voice Messaging
- ✓ Caller ID
- ✓ Call waiting
- ✓ Speaking clock

Types of phone calls

- ✓ Local call: telephone communication between two correspondents living in the same area and served by the same local exchange (applicable in fixed telephony)
- ✓ Long distance call: Telephone communication between two correspondents in two cities
- ✓ International call: Telephone communication between two correspondents from two countries
  - o Landline telephone to landline the phone
  - o Landline to mobile telephone
  - o Mobile telephone to landline telephone
  - o Mobile telephone to mobile telephone

Phases of a call via a fixed telephone set[145]

- ✓ Preselection
  - o Handset off – hook by the caller
  - o Detection of the connection request / connection by the switch
  - o Sending of dial tone to the caller by the switch
- ✓ Recording and translation
  - o Dialing of the telephone number by the caller
  - o Decoding and storage of numbers by the recorder of the switch (recording)
  - o Determination with the routing tables of the switch the call routing (translation)
- ✓ Selection
  - o Transmission by switch A of the signaling necessary for the call to switch B

---

144 Téléphonie - IUT de Nice
iutsa.unice.fr/~frati/telephonie_DUT/Telephonie-2008-2009-Print.pdf

145 Réseau téléphonique : du RTC au RNIS Large Bande - Patrice KADIONIK
ftp://ftp-developpez.com/kadionik/reseau/reseau-telephonique.pdf

- o Analysis of the number and identification of the subscriber called by the switch B (available subscriber, subscriber already in communication, impossibility to establish communication)
- o Availability of the called party: sending of a signal to switch A to indicate the progress of the call
- o Reservation of a connection and activation of the ringing of the called party by the switch B
- o Sending a tone to switch A by switch B
- o Impossibility to establish a call: sending of signaling from switch B to switch A
- o Generation of busy tone and release of the reserved line from switch A

✓ Connection

- o Establishment of connection between the caller and the called party
- o Two-way transmission of exchanges between the two correspondents

✓ Taxation

- o Start of charging the call to the caller (telephone in the off – hook condition)

✓ Supervision

- o Signal quality monitoring
- o Failure detection

✓ End of the communication

- o Disconnection of subscribers
- o Release of resources used for communication

Establishment of an international call

2 options

✓ Option 1: Direct interconnection between the two telephone operators

✓ Option 2: Routing of the call via a carrier

- o Players : Originating operator + Carriers + Terminating operator

Time of a telephone call

3 times

✓ Connection time (Time required to establish communication between subscribers)

✓ Duration of the conversation (Time elapsed during the exchanges)

✓ Disconnection time (Time required to disconnect subscribers)

Necessary elements of the telephone service

3 elements

✓ Switches: Organs managing telephone connections (intelligence for establishing communications)

✓ Subscriber Terminals: telephones, fax, modem, PDA

✓ Transmission media (links): Wired links (metallic cable or optical fiber) or wireless links (radio links or satellite links) between the various central offices or switches and subscribers

Billing of telephone services

✓ Prepaid services (consumer account debited)

✓ Postpaid services (services billed after use)

✓ Subscription (access to a monthly usage level against payment)

✓ Special plan (offers subject to conditions)

Network of a telephone operator

✓ Transmission network

- o Transport of voice
- o Costitution : multiplexers and links
- o Link technologies: optical fiber, microwave links and coaxial cable

✓ Switching network

- o Switching of traffic between the caller and the called party
- o Constitution : set of switches
- o Transfer by the network of transmission of traffic to the switch via input ports

Functions of a conventional telephone network

3 basic functions

✓ *Subscribers interconnection*

- o Connection of subscribers and transport of information via dedicated support for the duration of the communication

- ✓ *Signaling*
  - o Exchange of information necessary for access, call and connection
  - o Exchange of messages or frequency signals for the establishment, the breaking of the communication and its support (on the basis of the numbering)
- ✓ *Operation*
  - o Exchange of information and commands (messages) for network management (traffic measurement, commissioning, etc.)[146]

Parts of a conventional telephone network

- ✓ *Distribution:* Part of the network connecting the telephone subscribers to the nearest switch by different means and technologies
- ✓ *Switching:* Intelligent part of the network connecting telephone subscribers by allocating temporary circuits
- ✓ *Transmission:* Set of links connecting the network switches (radio links and fiber optics)

Functions of telephone switching

Processing of all real-time connection requests

- ✓ Referral of communications
- ✓ Concentration of telephone traffic
- ✓ Subscriber taxation
- ✓ Connection functions (connection of the caller to the called party by allocating a transmission resource)
- ✓ Call processing and signaling functions (processing of connection requests and services)
- ✓ Administration functions[147]
- ✓ Transfer of traffic to another switch via the transmission network
- ✓ Communication monitoring

Telephone network and its sub-networks

Different sub -networks of a telephone network

4 subnetworks

Access network

- o Connection of terminal equipment to the switching network
- o Analog or digital access (ISDN, xDSL, leased line, etc.)
- o Wired or wireless links

Signaling network

- o Network responsible for carrying signaling information (data) within the telecommunications network

Intelligent Network

- o Network operated for the provision of voice-oriented value-added services such as toll free, televot, billing, prepaid, etc.
- o Network consisting of application servers containing service logic (programs) and service data

Management Network

3 constituent elements

- o EMS (Element Management System): equipment management system provided by the telecoms manufacturer for the operation of equipment
- o OSS (Operation Support System): Network Information System (Network and Technical Service Management)
- o BSS (Business Support System): commercial information system (management of commercial and customer services[148]

Criteria for the evaluation of a telephone network

3 main criteria

- ✓ Quality of service: Reliability of the service provided
- ✓ Capacity: Number of communications supported

146 Trafic et performances des réseaux de télécoms - Georges Fiche et Gérard Hébuterne

147 Réseaux d'accès : du Réseau Téléphonique Commuté à la fibre optique, Telecom ParisTech http://perso.telecom-paristech.fr/~coupecho/cours/reseauxdacces.pdf

148 Réseaux et Services de Télécommunication Concepts, Principes et Architectures http://efort.com/r_tutoriels/ARCHITECTURES_EFORT.pdf

simultaneously

- ✓ Coverage: range of the service (Extent of the territory served)

Performance criteria of a telephone network

- ✓ Network availability (probability of placing a new call)
- ✓ Maintaining of communication (Cut - off rate)
- ✓ Network coverage (received signal level)
- ✓ Network capacity (quantity of simultaneous communications)
- ✓ Congestion rate (connection request percentage not established)

IP TELEPHONY

- ✓ All telephone services using an IP infrastructure as a means of transport
- ✓ Digitization of the human voice and transmission by TCP/IP as a data packet

Voice Over IP (Voice Over IP)

- ✓ A set of protocols for transporting voice communications over an IP data network (private network or the Internet)
- ✓ Alternative to traditional telephone network infrastructure to carry the human voice

Voice over the Internet

- ✓ Transmission of the human voice via the public Internet network

Equipments needed for voice over IP

- ✓ *Router*: Essential element providing of the routing of packets
- ✓ *Gateway*: Interconnection between the IP network and the PSTN (coding, decoding packetization of the voice)
- ✓ *Gatekeeper*: Authentication, authorization and supervision of calls (conversion of telephone number to IP address and vice versa)
- ✓ *Administration Server*: Billing of post or prepaid customers through the collection of CDRs.[149]

Operations of IP telephony

- ✓ Digitization of voice and packetization
- ✓ Routing of packets through routers and servers to the destination
- ✓ Reclassification of packets by sequence number
- ✓ Conversion of packets to voices at the reception

Packet of data

- ✓ Packet of data: 4 elements
  - o A header indicating the source and the destination
  - o Sequence number
  - o Data block
  - o Error verification code

Types of IP telephony

3 types of IP telephony

1.- IP Telephony from Computer to Computer (PC to PC)

- ✓ Call between two computers via the Internet
- ✓ Call between two computers via intranet or extranet

Elements involved in the scenario between 2 computers

- ✓ A computer at each end
- ✓ A modem at each end
- ✓ Connection on both sides to the PSTN (Switched Telephone Network)
- ✓ Connection of the PSTN to the Internet on both sides by an ISP
- ✓ Internet (center)

2.- IP Telephony from computer to Telephone

- ✓ Call between a computer and a telephone

Elements involved in the scenario between a computer and a telephone

2 categories of link

On the computer side

- ✓ A Modem

149 La Voix sur le Réseau IP https://www.itu.int/ITU-D/finance/work-cost-tariffs/events/tariff-seminars/cameroon-04/abosse_voix_sur_le_reseau_ip_fr_final.pdf

- ✓ Connection to PSTN (Switched Telephone Network)
- ✓ Connection of the PSTN to the Internet by an ISP

On the the telephone side

- ✓ Connection to PSTN (Switched Telephone Network)
- ✓ A gateway

3.- IP telephony from telephone to telephone

For communication between 2 IP telephones

Elements involved in the scenario between 2 telephones

- ✓ Telephone set on both sides
- ✓ PBX on both sides
- ✓ Gateway on both sides
- ✓ Connection of the gateway on both sides to the Internet or intranet or extranet

Voice over IP softwares

2 categories: free software and proprietary softwares

- ✓ Free software: Ekiga, Kphone, Linghone, Wengophone, etc.,
- ✓ Proprietary software: Microsoft netmeeting, Teamspeak, Skype, Google talk, Windows live messenger, etc.

Comparison between traditional telephony and IP telephony

Comparison criteria	Classic telephony	IP telephony
Commutation	Circuit	packet
Time	Real	delayed
Bandwidth	Inefficient usage	Better use
Services	Limited number	New services
Equipment costs	High	Low
Quality of service	Good	Less good

PABX (Private Automatic Branch Exchange)

- ✓ Private Branch Office
- ✓ Private telephone switch used in companies
- ✓ Switching of calls within an organization
  - o Management of incoming and outgoing calls
  - o Establishment of communication / telephone calls between offices within a company

Main functions of a PABX

- ✓ Connection of company's internal extensions
- ✓ Connection of internal extensions to the PSTN

Types of PABX solutions

- ✓ PABX (conventional PABX)
  - o Routing of calls by physical switches
- ✓ Virtual PBX (PBX Hosted on a Remote Server)
- ✓ IPBX (PBX-VoIP): software PBX system
  - o Routing of calls over the Internet
- ✓ Virtual IPBX (IPBX Hosted on a Remote Server or Offline PBX Service)

Services of a PABX

- ✓ *Abbreviated dialing*: Abbreviated dialing code for certain numbers
- ✓ *Callback*: Auto redial test for any busy line called
- ✓ *Line identification*: On-screen display of the caller's number
- ✓ *Conference call*: Multiple telephone call (Addition of other subscribers in the same telephone conversation)
- ✓ *Speakerphone (hands-free operations)*: Ability to speak without grasping the handset thanks to the microphone and loudspeaker
- ✓ *Message indicator*: Indication of reception of a message
- ✓ *Last number redial*: Operation facilitated by press-

ing a dedicated key

- ✓ *Voicemail*: Recording of messages (messages related to missed calls)
- ✓ *Transfer of calls*: Possibility of transferring a call to another number[150]

CELLULAR TELEPHONY

Mobile telephony or cellular telephony

- ✓ Wireless telephone communication means
- ✓ Telecommunications infrastructure allowing to use mobile or cellular telephones

Telecommunication system meeting the constraints of subscriber mobility in the network

- o Motivation: Electronic Communications on the Move (Subscriber Mobility)
- o Technical means: Wireless communications
- o Transmission medium: radio waves (Electromagnetic waves)

General information on cellular telephones

- ✓ Access to telephony and data services on the move (everywhere)
- ✓ Connection of the subscriber to the network by radio waves
- ✓ Territory to be served cut into cells of different dimensions
- ✓ Cell: area covered by a base station (BTS)
- ✓ BTS: Telecommunications infrastructure (transmitters / receivers, antennas, cables, etc.) installed in different locations
- ✓ Management by the cell of subscribers in its coverage area
- ✓ Linking of each base station to the switch via the Base Station Controller (BSC)
- ✓ Automatic transfer of communications from one cell to another to ensure continuity of communications (user mobility, signal loss, interference, etc.)
- ✓ Use by each mobile telephone of a separate and temporary radio channel to communicate with the BTS

Foundation of cellular telephony

Radiotelephony and Cell: Continuous coverage of a geographical area with base stations connecting the terminals with electromagnetic waves

- ✓ *Radiotelephony*: Communication techniques (human voice, text) supported by wireless links
- ✓ *Radiotelephony*: Radio telephony with mobile devices
- ✓ *Cell*: geographical service area of the cellular network covered by a base station
  - o Scope of a cell : from a few meters to a tens of kilometers

Functions of the cellular network

Same basic functions as the basic telephone network

- ✓ Subscribers interconnection
- ✓ Signaling
- ✓ Operation

Specific function of the cellular network

- ✓ Mobility management of telephone terminals

Operations of cellular network

- ✓ Request by the cellular subscriber for allocation of a frequency pair via the signaling channel
- ✓ Analysis of the numbers of the two correspondents by a central database
- ✓ After conclusive results, search for the requested subscriber in the database
- ✓ For a PSTN cable network subscriber, establishment of a link with the PSTN central office
- ✓ For a cellular network subscriber, BTS response in charge to the central computer ordering a connection through the central switch[151]

Frequencies of cellular telephone systems

Set of frequency bands allocated to cellular telephone systems

- ✓ GSM (2G) frequencies used throughout the world
  - o 8850 MHz, 900 MHz, 1800 MHz, 1900 MHz

150 Téléphonie numérique et téléphonie IP – David BENSOUSSAN

151 Généralités sur les réseaux cellulaires - http://www.memoireonline.com/03/12/5461/Interconnexion-entre-deux-reseaux-cellulaires-des-normes-GSM-par-faisceau-hertziens-cas-de-CCT-et.html

- ✓ UMTS (3G) frequencies used throughout the world
  - o 7700 MHz, 800 MHz, 850 MHz, 900 MHz, 1500 MHz, 1800 MHz, 1900 MHz, 2100 MHz, 2600 MHz, 3500 MHz
- ✓ LTE (4G) frequencies used throughout the world
  - o 7700MHz, 800MHz, 900MHz, 1700MHz, 1800MHz, 1900MHz, 2100MHz, 2300MHz, 2500MHz, 2600MHZ
- ✓ One frequency pair for each communication
  - o One frequency for cellular telephone connection - BTS (Uplink)
  - o One Frequency for BTS Link - Cellular Phone (Downlink)
- ✓ Use of other frequency bands (GHz band) for BTS links -Switch via the BSC

Access to mobile telephony

- ✓ Compatible mobile telephone connected to the cellular network
- ✓ Digital tablets with call capability

Types of Subscriber Terminal

3 Types

1.- Portable
- ✓ Most popular, miniaturized and lightweight
- ✓ Power: 0.6 watt

2.- Transportable terminal
- ✓ Transported by car
- ✓ High power (3 watts)
- ✓ Use of battery of the transport vehicle

3.- Fixed terminal
- ✓ Installed in offices
- ✓ Transmission power (3 watts)
- ✓ 110V or 220V power supply[152]

Characteristics of mobile networks

7 characteristics of any mobile telephone network

1.- Wireless links between the mobile and the system
- ✓ Freedom of movement of the subscriber
  - o Links established using electromagnetic waves

2.- Need to identify terminals
- ✓ Establishment of a mechanism for identifying the terminals connected to the network

3.- Need to locate subscribers
- ✓ Mechanism to determine at any time the location of a subscriber
  - o Mobility Management

4.- Need for a complex terminal
- ✓ Several complex functions to complete
  - o Telephone communication
  - o Transmission and reception of signals
  - o Power management
  - o Identification management
  - o Advanced processing capacity

5.-Use of a complex business model

3 options
- ✓ Mobile service providers
  - o Sale of services to subscribers
  - o Customer relationship management
  - o Global service management
- ✓ Mobile network operators
  - o Deployment of infrastructure and provision of services
- ✓ Virtual mobile network operators
  - o Capacity rental of a mobile network operator for serving customers

6.- Need for a specialized support service
- ✓ Specialized support service to set up for the management of information quests, orders, breakdowns and billing due to the complexity of the services provided

7.- Simple generic model of a mobile system
- ✓ Definition of terminal access mechanism to centralized network resources

152 Généralités sur les réseaux cellulaires - http://www.memoireonline.com/03/12/5461/Interconnexion-entre-deux-reseaux-cellulaires-des-normes-GSM-par-faisceau-hertziens-cas-de-CCT-et.html

o Multiple access techniques[153]

Media for cellular networks

- ✓ Wireless links between terminals and base stations
- ✓ Wired and wireless links between base stations and base station controllers

Main steps of a call via a cellular system

- ✓ Conversion of sound into electrical signal by the microphone
- ✓ Digitization of electrical signal by an analog to digital converter
- ✓ Transformation by the transmit antenna of the signals into electromagnetic waves (OEM)
- ✓ Sending of electromagnetic waves to the base station by the user's terminal
- ✓ Sending of electromagnetic waves to the MSC via the BSC by the BTS after signal processings
- ✓ Sending of electromagnetic waves to the BTS of the called party via the BSC by the MSC after various signal processings
- ✓ Transmission of the electromagnetic waves to the called party cellular telephone by the BTS
- ✓ Conversion of electromagnetic waves collected into electrical signal on the receive side
- ✓ Conversion of the electrical signal into sound message by the Loud speaker of the cellular telephone

From speech to radio transmission

- ✓ Transformation of sound waves into electrical signal
- ✓ Digitization of the electrical signal
- ✓ Source coding
- ✓ Channel coding
- ✓ Signal interleaving
- ✓ Encryption of the electrical signal
- ✓ Formatting of bursts
- ✓ Modulation of the electrical signal
- ✓ Transmission in free space

From radio transmission to speech

Reverse process

- ✓ Capture of the radio wave
- ✓ Demodulation of the electrical signal
- ✓ Reverse operation of formatting of bursts
- ✓ Decipherment
- ✓ Reverse operation of signal interleaving
- ✓ Channel decoding
- ✓ Source decoding
- ✓ Transformation of the electrical signal into sound

Process of a Telephone call over a cellular network

- ✓ Capture of the sound of the voice through the cellular telephone
- ✓ Conversion of the sound of the voice into an electrical signal (analog) by the microphone
- ✓ Digitization of the analog electrical signal by the terminal
- ✓ Transmission of the digitized signal in the form of electromagnetic waves by the telephone to the base station (BTS)
- ✓ Transfer of the signal by the BTS after procesing to the BSC
- ✓ Transfer of the signal by the BSC to the mobile switching center
- ✓ Search for the subscriber requested by the MSC
- ✓ Transfer of the call to the BTS serving the subscriber request via the home BSC
- ✓ Signal transmission to the phone by the BTS
- ✓ Conversion of electromagnetic waves by the antenna of the cellular telephone into electrical signal
- ✓ Conversion of the electrical signal in sound by the loudspeaker after processing

State of the cellular terminal

- ✓ Sleep mode: Terminal of the user connected at anytime and anywhere to the network through the beacon channel

153 Understanding Telecommunications Networks - ANDY VALDAR http://www.theiet.org/resources/books/telecom/19273.cfm

- ✓ In use: activation of the traffic channel (beacon channel maintained active)

Parts of the cellular telephone

2 parts: hardware and software

1.- Physical elements of a cellular telephone

- ✓ An Antenna
- ✓ A circuit board (circuit board: brain of the telephone)
- ✓ LCD screen (crystal display screen)
- ✓ A keyboard
- ✓ A Microphone
- ✓ A Speaker
- ✓ A Battery
- ✓ SIM card (except some types of CDMA telephones)

2- Softwares or operating systems of Cellular telephone

- ✓ Operating system: Softwares designed for the operation of the cellular telephone
- ✓ Roles of the operating system: Interfacing between the user and the hardware
    - o boot, access to available services, shutdown
    - o operation of the various hardware peripherals
- ✓ Main operating systems: Android, iOS, Windows phone, Black Berry OS, Symbian, Limo, Bada, Meego, Palm webos.

Features of the cellular telephone

- ✓ Link with the base station: electromagnetic waves (Radio waves)
- ✓ Interface with the user: signal level indicator on the screen
- ✓ Energy autonomy: removable or integrated battery

SIM card

- ✓ SIM card (Subscriber Identity Module): Subscriber Identity Card
- ✓ Electronic chip containing information about the user, services subscribed and the operator
- ✓ User accessible element
- ✓ Achievement of all the functionalities necessary for the transmission and the management of displacements
- ✓ SIM card function: storage and management of a series of information
- ✓ SIM: mini-database

SIM card data

- ✓ Administrative data
- ✓ Security related data
- ✓ User data
- ✓ Roaming data
- ✓ Network data[154]

Cellular telephone Uses

Main uses of the cellular telephone

- ✓ Telephone call
- ✓ Voice Messaging
- ✓ Sending/reception of SMS/MMS
- ✓ Sending/reception of e-mail (e-mail)
- ✓ Connections to social networks (sending / reception of texts, photos and videos)
- ✓ Web surfing
- ✓ Instant messaging
- ✓ Research on the Internet
- ✓ Online banking
- ✓ Downloading of applications
- ✓ Geolocation

Other possible uses of the smart phone

- ✓ Camera (pictures and recording of videos)
- ✓ Watch
- ✓ Alarm
- ✓ Calculator
- ✓ Agenda

---

154 Principes de base du fonctionnement du réseau gsm - cédric demoulin, Marc Van Droogenbroeck https://orbi.ulg.ac.be/bitstream/2268/1381/1/demoulin2004principes.pdf

- ✓ Storage of contact information (address book)
- ✓ To do-list
- ✓ Appointment Monitoring and Reminder Settings
- ✓ Games
- ✓ PDA
- ✓ Mp3 player
- ✓ Etc.

Smartphone telephone (smartphone)

- ✓ Mobile phone with touch screen, digital camera, PDA and some laptop functions[155]
- ✓ Data entry by keyboard or touch screen

Criteria of choice of a cellular telephone

Several criteria to check before choosing a cellular telephone

1.- Standard used by the terminal

- ✓ Standard: indication of the services accessible on the terminal
- ✓ GSM (Groupe Special Mobile now Global System for Mobile Communication, 2G): the first standard for digital cellular telephony
- ✓ 2G: Telephony and SMS
- ✓ GPRS (General Packet Radio Service, 2.5G): Telephony, SMS, Low-speed Internet
- ✓ EDGE (Enhanced Data Rate for GSM Evolution, 2.75 G): Telephony, SMS / MMS, enhanced throughput
- ✓ 3G: All basic services and broadband Internet
- ✓ 3.5G and 3.75G: All basic services and Internet with improved flow in both directions
- ✓ 4G: All basic services and Internet with high-speed video support, mobile TV, real-time multimedia, online games, etc.

2.- Frequency bands used

- ✓ Frequency bands allocated to cellular communications: 800MHz, 900MHz, 1800MHz, 1900MHz, 2100MHz and 2500 MHz
- ✓ Dual Band Phones: 2 frequency bands
- ✓ Tri-band phones (tri-band): 3 frequency bands
- ✓ Quad-band phones: 4 frequency bands
- ✓ Peta phones: 5 frequency bands
- ✓ Multiple Frequency Bands: Possibility for the terminal to be used on multiple cellular networks

3.- Level of sensitivity of the terminal

- ✓ Low signal detection capability
- ✓ Necessary in areas poorly covered by radio waves
- ✓ Very sensitive telephones, more adapted to environments poorly covered by radio signals

4.- Size and type of terminal screen

- ✓ Importance of a screen adapted for the data service
- ✓ Navigation provided by large touch screens
- ✓ Screen suitable for reading, writing messages, and browsing the Internet
- ✓ Monochrome or colors screens

5.- General ergonomics of the terminal

- ✓ Aesthetic object and easily manipulated
- ✓ Menu layout, key locations
- ✓ Ease of navigation.

6.- Terminal functions

- ✓ Quantity of services available on the terminal
- ✓ Available functions: telephony, SMS / MMS, videophone, voice recognition, ringtones or logo downloads, games, internet access, geolocation, etc.

7.- Energy autonomy

- ✓ Terminals charged by removable or built-in batteries
- ✓ Energy Autonomy: Big challenge of cellular terminals
- ✓ Autonomy from a few hours to a day
- ✓ Services + applications = high consumption of electrical energy
- ✓ Greater energy autonomy = deactivation of certain applications

155 Smartphone
https://fr.wikipedia.org/wiki/Smartphone

8.- *Weight and volume*

- ✓ Important factors in the choice a cellular telephone
- ✓ Consumer tastes: Sophisticated, lightweight and low volume phone

9.- *Terminal price*

- ✓ Determining factor in the choice of a terminal
- ✓ Higher quality and options for an expensive terminal
- ✓ Subsidy of expensive phones by telephone operators for subscriber attraction[156]

Elements of a second-generation network (2G)

- ✓ BTS: Base Transceiver Station or Base Station
- ✓ BSC : Base Station Controller:
- ✓ MSC: Mobile Switching Center
- ✓ HLR: Home Location Register
- ✓ VLR: Visitor Location Recorder
- ✓ AUC: Authentication Center
- ✓ EIR: Equipment Identity Register
- ✓ GMSC: Gateway of the Mobile Switching Center)

Functions of 2G network elements

Specific functions for each element

Functions of a BTS

- ✓ Communication between the mobile subscriber and the cellular network
- ✓ Transmission and reception of radio signals from the terminal
- ✓ Transmission and reception of network radio signals (BSC)
- ✓ Interface between the terminal and the system
- ✓ Assignment of communication channels to mobiles
- ✓ Permanent transmission of signaling

Considerations in the choice of a BTS

- ✓ Number of potential users in the area
- ✓ Terrain configuration (geographical relief, presence of buildings)
- ✓ Nature and density of buildings (houses, buildings, concrete buildings, ...)
- ✓ Location (rural, suburban, or urban)[157]

Functions of BSC

- ✓ Management of BTS
- ✓ Traffic and signaling switching between BTS and MSC
- ✓ Control of call transfers executed by BTS
- ✓ Frequency allocation and power control

Functions of Mobile Switching Center (MSC)

- ✓ Registration of users
- ✓ Mobile subscribers search
- ✓ Switching of communications between the calling and called subscriber
- ✓ Establishment of calls within the same MSC
- ✓ Establishment of communications between a mobile and another MSC
- ✓ Interconnection between the mobile network and other networks (fixed and mobile)
- ✓ Call monitoring
- ✓ Management of the radio channel change procedure during communications, automatic load balancing
- ✓ Transmission of short messages
- ✓ Call transfers from cell to cell
- ✓ Monitoring data communications
- ✓ Control of cellular sites
- ✓ Billing
- ✓ Maintenance

Functions of the HLR (Home Location Register)

Primary database attached to the MSC

156 La téléphonie mobile : technologies, acteurs et usages par M. Benjamin Savoure http://junon.u-3mrs.fr/u3ired01/Main%20docu/telecom/mem-savoure.pdf#sthash.0cjfmV5d.dpuf

157 Chapitre 1 : Généralités sur les réseaux cellulaires - http://www.memoireonline.com

- ✓ Data management of network subscribers
  - o Management of the location of network subscribers
- ✓ Storage of information
  - o Characteristic data o a subscriber
  - o Storage of static records (subscriptions, subscribed options and additional services available)

Functions of VLR (Visitor Location Register)

- ✓ Secondary database attached to the MSC
  - o Management of the location of mobile subscribers in their area (MSC service area)
  - o Storage of mobile subscriber movement information in a location area
  - o Decrease of the HLR load
- ✓ Additional information stored in the VLR
  - o Identity of the location area
  - o Temporary identity of the mobile subscriber
  - o Mobile number of Roaming station
  - o Terminal status (busy / free / no answer, etc.)

Functions of the AUC (Authentication Center)

- ✓ Data storage for each mobile subscriber
  - o IMSI authentication
  - o Encryption of communication on the radio link

Functions of the EIR (Equipment Identity Register)

- ✓ Database comprising security and identification information related to a cellular telephone
- ✓ Storage of the IMEI of the telephone
- ✓ Classification of terminals
  - o Whitelist
  - o Gray list
  - o Blacklist
  - o Unknown terminals by the EIR

Functions of Gateway Mobile Switching Center (GMSC)

- ✓ Gateway for interconnection with other fixed and mobile telephone networks

"Mobility of Services" in Cellular telephony Networks

- ✓ Mobility in communication networks: ability to access, from anywhere, all services normally available in a fixed and wired environment[158]
- ✓ Objectives of mobility
  - o Allowing users to have telecoms services (transmission / reception) on a coverage area
  - o Pursuing a communication while moving[159]
- ✓ Basis of mobility: call transfer from cell to cell
  - o Handover or Handoff: Transfer of communications in progress between cells (mobile in use and mobile on)

Means: monitoring of the signal level by the network to decide whether to transfer the communication or not

- ✓ Intercellular transfer: transfer of the communication to another cell (consequence of the mobility of the user)

Handover functions

- ✓ Allowing users to move during a call (from sector to sector or from cell to cell)
- ✓ Avoiding permanently the break of the communication link

Generations of cellular networks

5 generations of cellular networks

First generation (1G)

First Generation of Cellular System (1G): Analog Mobile Communications (1981)

- ✓ Standards for the first generation of cellular systems (1G)
  - o AMPS (*Advanced Mobile Phone System*), launched in the United States, analog network based on FDMA (Frequency Division Multiple Access) technology
  - o NMT (*Nordic Mobile Telephone*) designed pri-

158 Introduction aux Réseaux Mobiles – Présenté par: Samuel Pierre, Ing., Ph.D. Max Maurice, Ing.

159 Cours Architectures des réseaux mobiles - Gestion de la mobilité – Dept. Télécoms - INSA Lyon Fabrice Valois, Laboratoire CITI

marily in the Nordic countries and used in other parts of the planet

- o TACS (*Total Access Communications System*), a network based on AMPS technology, and heavily used in the UK

1G Services

- ✓ Only Voice Service (Frequency Multiplexing: FDMA)

Limits of 1G

- ✓ Analog radiotelephones: installed in cars or carried in suitcases
- ✓ Poor voice quality
- ✓ Poor battery quality
- ✓ Huge and highly energy-consuming terminals
- ✓ Limited service coverage
- ✓ Absence of confidentiality in communications
- ✓ Weak reliability of the handoff

Second Generation (2G)

- ✓ Second generation of mobile networks (2G): Breaking with the first generation of cell phones through the transition from analog to digital (1991)
- ✓ 2G mobile telephone standards: TDMA and CDMA
- ✓ TDMA: Time Division Multiple Access (TDMA)
  - o GSM (Groupe Special Mobile became *Global System for Mobile communications*), most used standard in Europe in the late twentieth century, supported in the US
  - o Use of 900 MHz and 1800 MHz frequency bands in Europe.
  - o Use of the 850 MHz and 1900 MHz frequency bands in the United States
  - o IDEN: Proprietary network used in the United States by Nextel
  - o PDC: Standard used exclusively in Japan
- ✓ CDMA : Code Division Multiple Access (CDMA)
  - o IS - 95: Known as CDMA used in the United States and some parts in Asia

Services of 2G

- ✓ Voice (telephone communication)
- ✓ SMS (Short message service: 160 characters)

Characteristics of 2G

- ✓ Confidentiality of telephone communications

9.6kb / s rate for the voice

Limitations of 2G

- ✓ Operation of the terminal based on a strong signal
- ✓ Inability to provide heavy data service (video for example)

Transition from 2G to 2.5G (GSM to GPRS)

- o Strategy: New IP core network

2.5G (GPRS)

- ✓ GPRS: General Packet Radio Service (1997)
- ✓ Intermediate step between 2G and 3G
- ✓ Goal: Data transmission over 2G networks
- ✓ Packet switching for data transmission

2.5G Services

- ✓ Digital telephony on the GSM network
- ✓ SMS (broadcast)
- ✓ MMS (Multimedia Messaging Service)
- ✓ Internet on the GPRS network
- ✓ Speed : 50kb/s to 144kb/s
- ✓ Push to talk over cellular (PoC)
- ✓ Instant messaging
- ✓ Internet applications for smart terminals via wireless application protocol (WAP)
- ✓ Point-to-Point Link(P2P): Inter-networking with the Internet (IP)
- ✓ Point-to-Multipoint Link (P2M)

2.5G network architecture
- ✓ Conservation of the 2G network architecture
- ✓ Addition of some other elements
    - o Elements dedicated to data communications

Transition from 2.5G to 2.75G (GPRS to EDGE)
- o Upgrade of the core network and the access network to reach speeds of 384 Kb /s
- o Conservation of GSM frequencies

2.75G (EDGE)
- ✓ EDGE: Enhanced Date Rates for GSM Evolution (1999)
- ✓ Step between 2.5G and 3G
- ✓ Goal: Improvement of data transmission throughput
- ✓ Packet switching

Services of the 2.75G
- ✓ Digital telephony on the GSM network
- ✓ SMS (SMS broadcast)
- ✓ MMS (Multimedia Messaging Service)
- ✓ Internet on the EDGE network
- ✓ Throughput: up to 384k /s

2.75G Network Architecture
- ✓ Conservation of the 2.5G network architecture
- ✓ Strengthening of data transmission capacity

Transition from 2.75G to 3G (EDGE to UMTS)
- o New frequency bands and improvement of the core and access networks

Third Generation (3G)
- ✓ UMTS: Universal Mobile Telecommunications Systems
- ✓ Designed and developed for mobile data services in mobile mode

3G services
- ✓ Basic services (Telephony, SMS / MMS, Internet)
- ✓ High-speed Internet access from a mobile device or computer
- ✓ Video Calling
- ✓ Video messages
- ✓ Television
- ✓ Geolocation
- ✓ Multimedia (text, sound, images, videos)
- ✓ Web applications (information portal, wml)
- ✓ Bit rate: up to 2Mb/s

Disadvantages of 3G
- ✓ Cost of the 3G license
- ✓ Challenges in building the 3G infrastructure
- ✓ High bandwidth consumption (frequencies)
- ✓ High price of 3G terminals
- ✓ High energy consumption of 3G terminals

3G network architecture
- ✓ Conservation of some elements of 2G network
- ✓ Replacement of some elements
    - o BTS, BSC, etc.
- ✓ Addition of some other elements
    - o Elements dedicated to the transmission of data

3.5G (HSDPA)
- ✓ HSDPA: High Speed Downlink Packet Access
- ✓ HSDPA: Mobile phone protocol with download speeds greater than 3G
- ✓ Intermediate step between 3G and 4G for increased throughput
- ✓ WCDMA-based technology (Wideband - Code Multiple Division Access)

3.5G Services
- ✓ All services of 3G
- ✓ Speed: 8 - 10 Mb / s (in the downward direction)

Architecture of the 3.5G
- ✓ Conservation of the architecture of the 3G network
- ✓ Strengthening of data transmission capabilities

3.75G (HSUPA)

- ✓ HSUPA: High Speed Uplink Packet Access
- ✓ HSUPA: Mobile telephony protocol offering high speed in the upstream direction (Improvement of 3G)
- ✓ Intermediate step between 3.5G and 4G for increased speed

3.75G service
- ✓ All services of 3G
- ✓ Speed: 5.6Mb / s (in the upstream direction)
- ✓ Reduction of latency

3.75 G network architecture
- ✓ Conservation of the architecture of the 3G network
- ✓ Strengthening of data transmission capabilities

Fourth Generation (4G)
- ✓ Designed and developed for mobile video services
- ✓ Downloading of an 80MB movie in 40 seconds in an advanced 4G network
- ✓ Bitrate: up to 100 Mb/s (mobility) and higher speeds (1Gb/s for low mobility communications (pedestrians and fixed users)
- ✓ Secure all-IP broadband solution for wireless modems for laptops, smartphones and other mobile devices
- ✓ High quality of service and high security
- ✓ Services everywhere and always

4G Services
- ✓ All services of 3G (telephony, SMS / MMS, Internet)
- ✓ High-speed Internet access from a mobile device or computer
- ✓ Video Calling
- ✓ Video messages
- ✓ Fixed and mobile television
- ✓ IP telephony
- ✓ Online games
- ✓ Real-time multimedia services

Disadvantages of 4G
- ✓ Cost of the 4G license
- ✓ High bandwidth consumption (frequencies)
- ✓ High price of 4G terminals
- ✓ High energy consumption of 3G terminals
- ✓ Significant investment in the network

4G network architecture
- ✓ Conservation of some elements of the architecture of the 3G network
- ✓ Additions of some elements
- ✓ Strengthening of data transmission capabilities

Fifth Generation (5G)
- ✓ Technology under development in several countries of the world
- ✓ Designed to enable more broadband connections
- ✓ Commercialization planned for 2020
- ✓ Bitrate: from 1Gb/s

5G Services
- ✓ All 4G services
- ✓ Downloading of a movie in seconds
- ✓ Broadband Internet at speeds of 300 miles / hour (480km / hour)
- ✓ Global deployment: 7 trillion connections (10 connections per user via smartphones, tablets and other terminals)
- ✓ Interactive multimedia

Features of 5G
- ✓ 7.35 Gb/s or 940 MB/s, between a base station and a fixed terminal
- ✓ 1.17 Gb/s, ie 150 MB/s, between a base station and an on-board terminal in a vehicle traveling at a speed slightly greater than 100 km/h

Benefits of 5G
- ✓ Broadband (High speed)
- ✓ High capacity
- ✓ Large data broadcast in Gbps (Gigabits per second)

Some forecasts for 5G

- ✓ Super-fast network
- ✓ Monthly consumption of more than 50 GB per consumer
- ✓ Explosion of data traffic
- ✓ All in the clouds (cloud computing)
- ✓ Accelerated growth of connected terminals

Conclusion

- ✓ 1G: Technically limited in relation to quality and confidentiality
- ✓ 2G: dedicated to VOICE

Main objective of other generations (3G, 4G, 5G): DATA TRANSMISSION on mobile networks

- ✓ 3G: focused on high-speed DATA TRANSMISSION
- ✓ 4G: designed for transmission of high-speed VIDEOS
- ✓ 5G: Designed for Multiple Broadband Connections

Portability of the telephone number

Possibility for a telephone subscriber to keep his or her telephone number (s) in the following cases:

- ✓ *Portability between telephone operators* : Possibility for a telephone subscriber to keep his or her telephone number during a change of telephone operator
  - o Retention of the telephone number from a telephone operator A to an operator B.
- ✓ *Geographical portability* : possibility for a telephone subscriber to keep his or her numbers by changing their address (change of living area)
  - o A fixed telephone number assigned to a subscriber living in New York may be used by this subscriber in his new residential area in California
- ✓ *Service portability*: possibility for a telephone subscriber to keep the service or services associated with the telephone number by switching from the fixed service to the mobile service and vice versa with said telephone number

Benefits of phone number portability

- ✓ Freedom of movement for telephone subscribers
- ✓ Best services (thanks to the competition)
- ✓ Search for better costs for services
- ✓ Access to new services

Interconnection of Telecommunications Networks

Interconnection: Connection of two or more public telecommunications networks

Interconnection: Exchange of traffic (voice, SMS, national and international traffic) between the networks of two operators[160]

Interconnection: Establishment of physical facilities allowing two operators to communicate with each other and beyond their respective networks[161]

Principle of the interconnection of telecommunications networks

- ✓ Allowing any user of a public telecommunications network to establish communication with any user of another public telecommunications network, under the most favorable technical and economic conditions[162]

Rationale for the interconnection of telecommunications networks

- ✓ Regulatory obligation (universal service for all)
- ✓ Possibility for a subscriber of a telephone operator A to reacht the subscribers of a telephone operator B.
- ✓ Limited market share of a given telephone operator (other networks with other subscribers to be connected)
- ✓ Economics of telecommunications networks

Means of interconnection of telecommunications networks

- ✓ Establishment of a link by cables or radio between the switches of the two operators to interconnect
- ✓ Programming in the telephone dial plans of the

160 Présentation Tarifs Nedjma ARPT 12 Oct 04 – www.arpt.dz/fr/doc/actu/sem/communications/journee-etude/med-kaddour.pp

161 Lignes directrices sur l'interconnexion – Site de l'Instance Nationale des Télécommunications : GLOSSAIRE

162 Lignes directrices sur l'interconnexion – http://www.intt.tn/upload/txts/fr/decision_35_version_fr.pdf

two operators for immediate detection of the other operator's telephone numbers

- ✓ Termination after analysis of an initiated call from operator A to operator B

Aspects of the interconnection of telecommunications networks

- ✓ *Technical aspects*: Evaluation of the traffic to be exchanged between the operators in both directions with a view to the adequate dimensioning of the interconnection links, choice of telecommunications equipment (radio, transmission medium, etc.), fault resolution procedures
- ✓ *Regulatory aspects*: Processing of regulatory issues, legal clauses and possible appeals or claims for non-compliance with commitments made by telephone operators
- ✓ *Commercial aspects*: Acquisition of interconnection equipment and determination of interconnection rate
- ✓ *Interconnection rate*: amount to be paid to operator B by operator A for termination of traffic (call, SMS / MMS, etc.)

Sharing of revenues generated by interconnection traffic

- ✓ Use of the resources of a network or telecommunications networks of other operators for the termination of telecommunications traffic
- ✓ Obligation to pay part of the revenues generated by the interconnection traffic to the other operator for the use of its network

Control of interconnection by the Regulator

Protection of consumers and operators

- ✓ Quality of service
- ✓ Availability (without interruption)
- ✓ Affordable cost

Roaming in cellular telephony

- ✓ Possibility for a cellular telephone subscriber to place and receive calls, send and receive messages, or access other services, including data services of his network outside the geographical coverage of his telecommunications network through another telecommunications network
- ✓ Use of the same terminal and the same phone number outside the network coverage area to access electronic communications services
- ✓ Use of another cellular network in another region to access services provided by the originating operator
- ✓ Mobility of telecommunications services worldwide
- ✓ Possibility for a European cellular subscriber to use his or her cell phone and his telephone number in Asia, America, and Africa to access the services subscribed at his or her cellular telephone operator

Roaming rationale

- ✓ Access to the service anywhere in the world
- ✓ Local operator: limited coverage (local, regional or national coverage)
- ✓ Local operator: no coverage in other countries

Roaming types

Three types of roaming service provided worldwide

- ✓ *Regional* roaming: roaming provided by two operators operating in the same country
- ✓ *International* roaming: roaming provided bilaterally by two operators operating in two different countries
- ✓ *Interstandard Roaming* : Roaming provided by two cellular operators using two different technologies (for example: CDMA, GSM)

Operation of the roaming service

- ✓ Service available for all cellular subscribers
- ✓ Service subject to a bilateral agreement between the home operator and the host operator
- ✓ Service subscription or automatic activation outside the coverage area of the originating network

Roaming agreement between two telecommunications operators

- ✓ Agreement between two operators
    - o Contractual
    - o Commercial
    - o Financial
    - o Technical

- ✓ Deployment of telecommunications infrastructure

Roaming Benefits

- ✓ For the originating network
  - o Customer satisfaction (availability of service anywhere
  - o Indirect revenue generation (through host networks)
  - o Competitive advantages
  - o Less expenses on setting up infrastructure in other places
- ✓ For the host network
  - o More virtual subscribers
  - o Revenue Generation
  - o Competitive advantages
  - o Optimal use of network capacity
- ✓ For the subscriber
  - o One telephone everywhere
  - o Availability of the service anywhere

Mobile Virtual Network Operators (MVNO)

MVNO (Mobile Virtual Network Operator): Virtual telephony network operator

MVNO: Provider of mobile telecommunications services without a network

- ✓ No frequency spectrum concession from the State (no frequency assignment)
- ✓ No own network infrastructure
- ✓ Leasing of Network Operator Capability (MNO: Mobile Network Operator) for the provision of telecommunications services
- ✓ Resale of services under the brand of the virtual operator
- ✓ Accounts of subscribers domiciled exclusively at the MVNO
- ✓ Management of customer service, billing, marketing and sales services
- ✓ Examples: Lebara, Lycamobile, Ortel, China Unicom, mobile Virgin, Netzero

Speech synthesis

- ✓ *Speech Synthesis:* A set of devices, equipments or algorithms designed to automatically generate artificial speech
- ✓ *Speech synthesis*: Technology using a sound synthesis to read a text with an artificial voice
- ✓ *Speech synthesis:* Reproduction of the human voice from a combination of words
- ✓ *Speech synthesis:* Synthetic voice reading of a digital text
- ✓ *Speech synthesis:* Gateway between the written and the oral

Applications of speech synthesis in telecommunications

- ✓ Element at the base of machine-man communication
- ✓ Short oral answers to questions from telephone subscribers
- ✓ Reading of accounts (telephone account balances)
- ✓ Reading of e-mail and fax (fax)
- ✓ Reading of database and website
- ✓ Access to services for the visually impaired people

Word Recognition and Speech Recognition

- ✓ *Word Recognition*: Set of technologies allowing a machine to recognize the human voice
- ✓ *Word Recognition*: Comparison of repeated words with those stored in the machine for interaction
- ✓ *Word Recognition*: Analysis of a word or phrase picked up by a microphone to convert it into machine-readable text

Speech Recognition

- ✓ *Voice recognition*: Automatic recognition of the speaker
- ✓ *Voice recognition*: Recognition by a computer system of the elements or words of a voice message[163]

163 Futura tech http://www.futura-sciences.com/tech/definitions/informatique-reconnaissance-vocale-3958/

Applications of speech recognition in telecommunications

- ✓ Automatic translation of telephone conversations with a foreign language contact
- ✓ Information servers by telephone
- ✓ Key word recognition by a voice command system
- ✓ Search for information by voice on a computer, cell phone or digital tablet (Google Voice service)
- ✓ Dialing of a telephone number by voice
- ✓ Voice interactions between users and machines

Voice recognition applications

- ✓ Use as a voice signature
- ✓ Use of the telephone to speak to machines
- ✓ Command and control of remote devices

Telecommunications traffic

- ✓ Traffic: importance and frequency of communications over a telecommunications network (telephone calls, sending of messages, packets or frames, etc.)
- ✓ Traffic: Amount of telephone calls or data messages carried over a telecommunications network
- ✓ Traffic: set of signals exchanges for the management of communications within a telecommunications network
- ✓ Traffic: Volume of calls, SMS, e-mails, instant messages, images, videos exchanged during a given time on a network
- ✓ Traffic: set of telecommunications-related activities within the telecommunications network (similar to the flow of vehicles on a motorway, to airplanes in space, to boats in the sea, etc.)
- ✓ Traffic: ON - NET (Traffic generated by two customers of the same telecommunications network)
- ✓ Traffic: OFF NET (Traffic from Network A to Network B)

Traffic from the point of view of a telecommunications operator

- ✓ Traffic: Payload (set of connections to be established at the request of users) of a telecommunications network
- ✓ Volume of minutes, SMS exchanged between two telecommunications operators per month or per year

Types of telecommunications traffic

- ✓ Telephony traffic
- ✓ Data traffic (for example: Internet)

Telecommunications traffic units

- ✓ Erlang, MOU, CS, CCS and kbps

The following relationships in telephony

- ✓ 1 Erlang = 60 MOU (Minute of Usage)
- ✓ 1 Erlang = 3600 CS/hour (Call - second)
- ✓ 1 Erlang = 36 CCS/hour (Hundred Call - Second)
- ✓ 1 Erlang = 64 Kilobits/second (data traffic)

Importance of telecommunications traffic

- ✓ *Traffic*: The rationale or justification of telecommunications networks
- ✓ *Traffic data*: a tool needed to dimension switching and network transmission
- ✓ *Traffic*: source of income for telecommunications operators (telephone operators, internet service providers, etc.)

Telecommunications traffic concepts

- ✓ *Offered traffic*: All communication attempts (telephone calls, SMS, messages, etc.) transiting through a telecommunications network
- ✓ *Carried traffic*: Percentage of successful offered traffic (answered traffic + unanswered traffic from called party)
- ✓ *Lost traffic*: Percentage of offered traffic not completed
- ✓ *Incoming traffic*: Communications flow entering a telecommunications network (as part of an interconnection with other operators)
- ✓ *Outgoing traffic*: Communications flows leaving a telecommunications network (as part of an interconnection with other operators)
- ✓ *Internal traffic*: Communication flows between users of the same network (Callers and called: subscribers of the same network)

Conventional telephony traffic

- ✓ Number of call attempts per day

- ✓ Duration of a call
- ✓ Average duration of a call

Mobile telephony traffic

- ✓ Number of call attempts
- ✓ Duration of a call
- ✓ Average duration of a call
- ✓ Number of SMS / MMS sent
- ✓ User mobility
- ✓ Etc.

Internet traffic

- ✓ Average data traffic
- ✓ Average connection time
- ✓ Number of Internet sessions
- ✓ Number of uploads and downloads
- ✓ Number of emails sent
- ✓ Etc.

Examples of telephone traffic

- ✓ 0.7 Erlang: Optimized telephone line traffic (42 minutes per hour of usage)
- ✓ 0.03 Erlang: average traffic of a residential subscriber line (about 2 minutes of busy time)
- ✓ 0.6 Erlang: Traffic of a professional subscriber line (reception professions, telemarketing, telephone switchboards, etc.)
- ✓ 70 milliErlang: Traffic of 100 minutes for 24 hours
- ✓ 0.16 Erlang: 10 minutes on the phone during an observation period of one hour (for example from 8h - 9h or from 12h -13h)

Congestion of telecommunications networks

- ✓ *Congestion*: Congestion : bottleneck or blocking in a telecommunications network due to high demand for connections at busy hours (rush hour)
    - o Busy hour (rush hour): Maximum connections requests from users
- ✓ *Congestion*: Inability of the network to satisfy all customers at the same time
- ✓ *Congestion*: Inability to transmit and process requests (switching)
- ✓ Consequences of congestion: Inability to connect a percentage of the customers
- ✓ *Congestion*: Situation experienced during peak hours, holiday periods (new year, etc.) and in case of emergency caused by a disaster

Causes of congestion in telecommunication networks

2 causes of congestion

- ✓ Excessive demand during unpredictable period
- ✓ Insufficient dimensioning compared to the number of telephone subscribers

Solutions to congestion

- ✓ Appropriate dimensioning of the transmission network
- ✓ Appropriate dimensioning of the switching

Scenarios in a congested network

3 scenarios

- ✓ Busy tones or a message indicating the network's inability to meet the demand at the same time
- ✓ Queuing of the message for possible delivery according to specified parameters
- ✓ Message rejected, returned or lost

Behavior of telephone subscribers in case of telephone congestion

3 types of rush hour telephone subscriber behavior

- ✓ Type 1: Abandon after a first unsuccessful call attempt
- ✓ Type 2: Continuous attempts until the call establishment
- ✓ Type 3: Continuous attempts until a certain time

CHAPTER 6

SOUND AND TELEVISION BROADCASTING

Broadcasting : Promotion and Marketing Instrument
Remote transmission by electromagnetic waves (radio waves)

BROADCASTING

- ✓ Broadcasting (transmission of information in all directions) by radio waves (electromagnetic waves)

2 major branches of broadcasting

- ✓ Sound broadcasting (speech and sound broadcasting)
- ✓ TV broadcasting (broadcasting of images accompanied by sound and music)

Sound broadcasting

- ✓ One-way radiocommunication (in one direction only, ie, from the transmitter to the receiver) for the purpose of broadcasting programs intended for the public
- ✓ Broadcasting by radio waves
- ✓ Dissemination of programs by radio waves, accessible to the public by means of a suitable reception device
- ✓ Public broadcasting of programs (Radio, Television)
- ✓ Medium of information, education and entertainment

Video broadcasting or television

- ✓ Remote Transmission, through a cable or radio waves, transient images of fixed or moving objects, together with sound

Importance of broadcasting (sound and television) at the social level

- ✓ Information medium easy to use
- ✓ Mirror of society
- ✓ Participation in social life
- ✓ Social Development Lever

Importance of broadcasting (sound and television) at the political level

- ✓ Means of participation in political life
- ✓ Strengthening tool for democracy

Importance of broadcasting (sound and television) at the cultural level

- ✓ Promotion of local cultures
- ✓ Broadcasting of cultural values throughout the world

Importance of broadcasting (sound and television) at the economic level

- ✓ Promotion of services and national production
- ✓ Economic Development Lever

Types of sound broadcasting stations

4 types of radio station

1.- State Radio Station

- ✓ Official communication organ of a country

Activities of a State radio

- ✓ Communication of information to the general public

2.- Commercial Radio Station (Private Company)

- ✓ Broadcasting company holding a commercial license from the State (supported by for-profit advertisers)

Activities of a commercial radio station

- ✓ Broadcasting of commercial announcements
- ✓ Broadcasting of news and information

3.- Religious Radio Station

- ✓ Communication organ intended to broacast religious content

Activities of a religious radio station

- ✓ Broadcasting of religious content *strictly speaking* (masses, sermons, praises, prayers, songs, books reading) and secular information interpreted from a religious perspective[164]

4.- Community radio station (rural radio, cooperative radio, participative radio, free, alternative, popular, educational radio)

- ✓ independent, non-profit, community owned organ managed and supported by people from a given community[165]
- ✓ Radio station set up and for a geographical community.

---

164 Communication https://communication.revues.org/3826

165 La voix des sans-voix : la radio communautaire, vecteur de citoyenneté et catalyseur de développement en Afrique http://africultures.com/la-voix-des-sans-voix-la-radio-communautaire-vecteur-de-citoyennete-et-catalyseur-de-developpement-en-afrique-7104/

Activities of a community radio

- ✓ communication and facilitation tool to provide quality programming meeting the information, culture, education, development and entertainment needs of the community
- ✓ Broadcasting of information for the inhabitants of a given locality, in the languages and formats best adapted to the local context
- ✓ Mobilization of community radio stations to announce events

Means of broadcasting of sound

- ✓ Hertzian waves (radio waves) from a transmitter
- ✓ Internet (access through a website)

Organization of broadcasting in analog mode

3 functions for a single operator

- ✓ Content Producer
- ✓ Transport of signals to the broadcasting site (implementation and operation of the STL - Studio - Transmitter Link
- ✓ Operation of a broadcasting site for the broadcasting of signals

Organization of broadcasting in digital mode

3 functions for three different operators

- ✓ Producer of content by TV station (TV channel or content editor)
- ✓ Transport operator: Transport and multiplexing of signals
- ✓ Broadcasting Operator: Programs broadcasting for the public

Infrastructure of a radio station

- ✓ Radio transmitter and accessories
- ✓ Transmitting antennas
- ✓ Transmission lines

Signals modulation

- ✓ Transposition of a baseband signal in a high frequency band to facilitate its transmission in free space (via antennas)
- ✓ *Modulation*: Product of two signals (information signal or signal in baseband and carrier signal generated by the local oscillator of the transmitter)

Analog and digital modulations

Analog modulations

- o Amplitude modulation (AM)
- o Frequency modulation (FM)
- o Phase Modulation (PM)

Main Digital Modulations

- o Amplitude Shift Keying (ASK)
- o Frequency Shift Keying (FSK)
- o Phase Shift Keying (PSK)
- o Quadrature Amplitude modulation (QAM)

Uses of Modulation

- ✓ AM: Amplitude Modulation
  - o Monophonic broadcasting and telephony
- ✓ FM: Frequency Modulation
  - o Stereophonic broadcasting, television broadcasting, telephony
- ✓ PM: Phase Modulation
  - o Transmission of digital signals over telephone circuits, microwave links (microwave links), satellite links
- ✓ ASK: Amplitude Shift Keying
  - o Transmission in optical cables
- ✓ FSK: Frequency Shift Keying
  - o Transmission of voice over transmission lines, high frequency radio transmission, telemetry
- ✓ PSK: Phase Shift Keying
  - o Modem, IEEE 802.11b, Data Communication
- ✓ QAM: Amplitude modulation of two quadrature carriers
  - o Cable TV, Wireless LAN, Satellite, Cellular telephony

Modulations used in sound broadcasting

- ✓ AM (Amplitude Modulation): variation of the amplitude of the carrier signal by the modulating signal (information signal or baseband signal) to generate an amplitude modulated signal

- ✓ FM (Frequency Modulation): Variation of the frequency of the carrier signal by the frequency of the modulating signal (information signal or baseband signal)

Frequency bands of radio stations

Two main bands used worldwide

- ✓ 530 KHz - 1700 KHz for AM (Amplitude Modulation: Amplitude modulation
  - o 117 AM radio stations with 10 KHz bandwidth per radio station
- ✓ 88 MHz -108 MHz for FM (Frequency Modulation: Frequency modulation)
  - o 100 FM radio stations at 0.2 MHz (200 KHz) per radio station (0.2 MHz (200 KHz) bandwidth per radio station and 0.2 MHz (200 KHz) side separation (0.1 MHz) and left (0.1MHz) to avoid overlapping signals

Means of Access to the Sound Broadcasting Service

5 ways

- ✓ *Traditional radio receiver (radio set)*: Terminal equipment designed to receive AM and FM signals
- ✓ *Website*: Access to audio content of radio stations via websites
- ✓ *Cellular telephone*: Access to audio content through cellular phone equipped to capture FM signals
- ✓ *Audio now*: Platform allowing to dial from a cellular telephone a number assigned to a radio station in order to listen in certain areas (countries) not covered by the broadcasts
- ✓ *Radio application:* access to radio broadcasts via a radio application hosted by a provider

Types of broadcast

- ✓ *Live broadcast*: Content broadcasting in real time
- ✓ *Offline broadcast*: Content broadcasting after registration

Transmission in broadcasting

- ✓ Cable transmission: Cable television
- ✓ Hertzian waves: Terrestrial TV broadcast using radio waves
- ✓ Satellite transmission: Television broadcasting using satellites

Transmitter

Device capable of generating a signal from the captured information

- ✓ Device capable of capturing information, encoding this information in a signal, and transmitting this signal through a communication channel
- ✓ Collection of electronic devices or circuits capable of converting the signal of the source into a form suitable for transmission
  - o Example: Broadcasting transmitter

Receiver

- ✓ Device capable of receiving a signal transited through a communication channel, decoding the information of the said signal, and restoring it in its original form
- ✓ Collection of devices or electronic circuits capable of capturing transmitted signals from the transmission medium, and converting them into the original form understandable by humans.
  - o Examples: TV and radio receiver

Functions of a radio wave transmitter

- ✓ Signal production suitable for transmission
- ✓ Signal processing (coding, digitization, modulation)
- ✓ Linkage to the transmission medium (guided or unguided medium)

Some signal processing in transmitters

- ✓ Amplification
- ✓ Possible encryption
- ✓ Compression
- ✓ Modulation (in the case of a transmission in free space)
- ✓ Filtering

Characteristics of a broadcasting transmitter

- ✓ Transmission of power: power propagated in the *unguided* transmission medium
- ✓ Frequency of transmission: Frequency of operation of the transmitter
- ✓ Stability (Amplitude Modulation): Ability to stay within the defined spectral limits

Functions of a radio wave receiver

- ✓ Signal capture
- ✓ Signal amplification
- ✓ Decompression
- ✓ Demodulation (in the case of modulated signals on transmission)
- ✓ Decoding
- ✓ Restitution of the signals in the original form

Characteristics of a radio wave receiver

- ✓ *Selectivity*: Ability to select desired signal or signals from all received signals
- ✓ *Sensitivity*: Ability to detect and use weak signals at the receiver input.
- ✓ *Fidelity*: Receiver ability to restitute the original signal as emitted
- ✓ *Stability*: Receiver ability to remain tuned to the desired frequency.
- ✓ *Dynamic*: Ratio between the maximum permissible power ratio at the receiver input and the minimum power required for its operation (power threshold)

OPERATION OF A RADIO AND TELEVISION STATION

3 fundamental elements for broadcasting by radio waves

1. Studio

- ✓ Production of content intended for the general public
- ✓ Conversion of content into electrical signals
- ✓ Transmission with a low-power transmitter the signals to the broadcasting site

2.-STL (Studio Transmitter Link)

- ✓ Transport of signals produced by the studio to the broadcasting site via a radio link
- ✓ Use of a microwave link (high frequency radio links) point-to-point for the transmission of signals to the transmitting site

3.-Broadcasting sites

- ✓ Reception of signals carried by the STL
- ✓ Processing of received signals (amplification, change of frequency, etc.)
- ✓ Broadcasting of the signals processed on the frequency of transmission of the radio station and the television channel on a hight range
- ✓ Coverage of a large area thanks to the altitude of the transmission site

Challenges of analogue sound broadcasting

- ✓ Limited radio coverage
- ✓ Saturation of the frequency band reserved for this telecommunications service (particularly FM)
- ✓ Radio interference with other signal sources

Criteria for choosing a modulation

- ✓ Performance (signal quality, radio coverage)
- ✓ Spectral occupancy (frequency spectrum congestion)
- ✓ Complexity of transmitters and receivers
- ✓ Energy consumption

Sound broadcasting services

- ✓ Broadcasting of speech and sound (broadcasting audio signals)
- ✓ RDS (radio data system): Digital data transmission in parallel of FM radio audio signals

Digital broadcasting

- ✓ Broadcasting of a binary signal, that is to say only composed of a succession of "0" and "1" on frequency bands (band III and band L mainly) different from those used for the FM (band II)[166]

Forms of Digital Broadcasting

- ✓ Digital Terrestrial Radio (DTR)
- ✓ Digital World Radio (DRM)
- ✓ Digital Radio via the Internet (Web radio)
- ✓ Digital Satellite Radio
- ✓ Digital Cable Radio

166 CSA.fr - La radio numérique terrestre http://www.csa.fr/Radio/Autres-thematiques/La-radio-numerique-terrestre

Principle of the radio wave broadcasting of digital terrestrial radio

- ✓ Transmission of several radio services (signals) on a single channel (multiplexing)
- ✓ Digitization and compression of the signal for bandwidth optimization
- ✓ Broadcasting the signal by radio over different frequency bands
- ✓ Band III: Band reserved for the broadcasting of digital terrestrial radio

Broadcasting of digital terrestrial radio signals

- ✓ Real time broadcasting
- ✓ Recording and offline access (Podcast)

Benefits of Digital Terrestrial Radio

- ✓ Better audio quality
- ✓ No bandwidth constraint
- ✓ Better broadcasting
- ✓ Broadcasting of multiple radios on the same frequency
- ✓ Possibility of conveying associated information
- ✓ Good quality in mobile reception

Digital broadcasting standards

- ✓ Digital Audio Broadcasting (DAB): European standard for broadcasting in ultra-short wave (VHF, UHF) and SHF microwaves
  - o 2 Variants of DAB: DAB and T-DMB
- ✓ Digital World Radio (DRM): global standard for short, medium and long wave digital broadcasting
- ✓ Digital Video Broadcasting (DVB): basic standard for television, also applicable to sound broadcasting
- ✓ Satellite Digital Radio (SDR): standard for satellite broadcasting recognized by ETSI for Europe
- ✓ Satellite broadcasting: S - DMB

Digital Radio Frequency Bands

- ✓ 174 MHz - 240 MHz (Standard T -DMB - DAB / DAB +)
- ✓ 1452 MHz - 1492 MHz (Satellite transmission)

Modes of broadcasting of digital radio

- ✓ DAB (Digital Audio Broadcasting) : compression of digitized signals before any broadcasting
- ✓ Digital cable radio: service available via digital cable television
- ✓ Digital satellite radio: Broadcasting of digital radio signals by satellite
- ✓ Digital radio on the Internet: Two options
  - o Listening in real time (streaming)
  - o Delayed listening (listening after downloading)

Reception of digital terrestrial radio

- ✓ Digital Terrestrial radio compatible receiver
- ✓ Display of text, images and videos on a receiver's built-in screen

Digital terrestrial radio Services

- ✓ Sound
- ✓ Text via a compatible receiver
- ✓ Images via a compatible receiver
- ✓ Videos via a compatible receiver

Internet radio (Web radio)

Net radio, radio streaming e-radio, Webcasting, online radio

- ✓ Radio station broadcasting over the Internet using streaming technology (Streaming)
- ✓ Audio broadcast service transmitted over the Internet rather than over radio waves

Technology used

- ✓ Streaming playback
  - o Live access to audio content

Operation of the web radio

- ✓ Launch of the audio player
- ✓ Search for the audio stream by the player
- ✓ Sending of the audio stream to the sound card
- ✓ Sound production by the speaker

Establishment of a webradio

- ✓ Codec (Encoder/Decoder)
- ✓ Appropriate equipment

- ✓ Software
  - o Streaming playback
  - o Browser capable of playing audio streams
- ✓ Hosting

Broadcasting techniques of Webradio

3 models

- ✓ Server Client model
- ✓ Peer-to-Peer Model
- ✓ Multicast model

Internet radio access (web radio or netradio)

- ✓ Computer or digital tablet with sound card or smart phone
- ✓ Internet connection
- ✓ Speaker

Advantages of webradio

- ✓ Absence of constraints related to the operating license
- ✓ Absence of power constraint (coverage area not related to transmission power)
- ✓ No frequency constraint (not limited by bandwidth availability as for traditional stations)
- ✓ No geographical limit (global coverage through the Internet)
- ✓ Broadcasting of photos, texts and links via the website
- ✓ Interactivity
- ✓ Conversation space
- ✓ Low cost of implementation
- ✓ Access from any smart terminals
- ✓ Variety of programs

Disadvantages of webradio

- ✓ Bad sound quality
- ✓ Possibility of technical obstacles
- ✓ Poor connection
- ✓ Software interference
- ✓ Impossibility to access certain web radios by dial up

Television broadcasting (Television)

- ✓ Set of procedures and techniques implemented to remotely transmit and receive audio-visual and data sequences of a scene
- ✓ Set of techniques used to remotely transmit non-permanent images of fixed or moving objects
- ✓ Television signal: superposition of a luminance signal and various control and synchronization signals

Activities of a television channel

- ✓ Production of television programs
- ✓ Broadcasting of television programs

Principle of radio wave Television

- ✓ Capture of still and moving images by the camera
- ✓ Conversion of images into an electrical signal by the camera
- ✓ Realization of a modulation using a carrier
- ✓ Various signal processing
- ✓ Conversion of the electrical signal into electromagnetic waves and radiation by the transmitting antenna
- ✓ Electromagnetic wave capture and conversion to an electrical signal by a receiving antenna
- ✓ Various processings
- ✓ Conversion of the electrical signal into images by the receiver and display on the screen

Parts of an analog TV channel

3 basic parts

- ✓ Signal sources: production of the luminance signal (optical sources of video signals or electronic sources of video signals)
- ✓ *Video* equipment: recording and production equipment (tape recorders, video recorders, mixing, switching and electronic rigging devices)
- ✓ *Transmitter*: device designed for the transmission of television signals

Composition : 2 parts (image and sound)

- o Separate signal processing (sound and image)
- o Mixing of signals modulated by a diplexer for

their radiation in the free space by a single transmitting antenna

Modes of transmission of Television

- ✓ Radio waves or hertzian waves or electromagnetic waves
- ✓ cables; (metallic cables and optical fiber)
- ✓ Satellite
- ✓ Internet (IPTV)

Structure of a Television System

- ✓ Capture of images
- ✓ Transducer
- ✓ Transmitter
- ✓ Transmission
- ✓ Receiver
- ✓ Transducer
- ✓ Reproduction of images

Phase of a television system

- ✓ Production of television programs
- ✓ Transmission of television programs
- ✓ Reception of television programs

Types of television channels

- ✓ Generalist channel: television channel targeting all audiences (broadcasting of news and entertainment programs)
- ✓ Thematic channel: television channel dedicated to to a theme such as sport, religion, culture, politics, scientific research, etc.

Television Technologies

- ✓ Analog Technologies (Analog Television)
- ✓ Digital Technologies (Digital Television)

TV picture

- ✓ Image decomposed into a set of points called pixel (Picture element)
- ✓ Point-by-point image analysis
- ✓ Scanning of points one after the other

Definition or resolution of television

- ✓ Number of points or pixels displayed on the screen
- ✓ Product of the number of points according to the vertical by the number of points according to the horizontal
- ✓ 625 lines in Europe and 525 lines in North America and Japan
- ✓ The higher the resolution, the better quality of service

Black and white television

- ✓ Broadcasting of a monochrome image (a black and white image)
- ✓ Restitution of a signal in shades of gray, ranging from white to black, resulting from the coding of the luminous intensity

Color television

- ✓ Broadcasting of color images
- ✓ Based on the three primary colors: Red, Green and Blue
- ✓ Color Combination (Red, Green and Blue) = source of all existing colors in television
- ✓ Color production based on simultaneous scanning of the image by the camera three times
- ✓ Composition of the video signal: a luminance signal and a chrominance signal

Standards of analog television

Three standards used worldwide

NTSC: National Television Standards Committee (United States of America and Japan)

- ✓ Video encoding on 525 interlaced lines at a frame of 30 images per second

SECAM: Sequential color with memory (France)

- ✓ Video encoding on 625 interlaced lines with a frame rate of 25 images per second

PAL: Alternating Line Phase (Europe)

- ✓ Video encoding on 625 interlaced lines with a frame rate of 25 frames per second

Modes of access to television

6 modes of access to television

1.- Free – to -air Television

- ✓ Reception with a receiver of television signals broadcast in free space

- ✓ Most popular mode of access

2.- Television by ADSL

- ✓ Access to free and paid channels, either by a single modem installed between the telephone jack and the television, or by a modem and a decoder
- ✓ Connection of a modem to the telephone jack, and connection of the decoder to the television receiver
- ✓ Connection of the two boxes by an Ethernet cable or power line or Wi-fi

3.- Television by optical fiber

- ✓ Connection of the subscriber to the network by optical fiber (access to free and paid channels)
- ✓ High definition television available on multiple television receivers simultaneously

4.- Cable Television

- ✓ Connection of subscribers to the network by cable
- ✓ Access to service facilitated by a digital decoder
- ✓ HD television available by cable

5.- Mobile TV

- ✓ Viewing of TV programs on EDGE and 3G networks
- ✓ QVGA screen and H264 file playback
- ✓ Access to live television

6.- Satellite TV provided by ISPs

- ✓ Access to free and paid channels and Video on demand
- ✓ Access facilitated by an antenna and a hybrid decoder (ADSL / Satellite)[167]

Means of access to television

Terrestrial television

- ✓ Indoor (single) or outdoor television antenna (Yagi antenna)
- ✓ Classic television receiver

Cable Network Operator

- ✓ Cable television service subscription
- ✓ Classic television receiver

Satellite television

- ✓ Satellite dish specially oriented towards one or more satellites
- ✓ Adapted analog or digital decoder
- ✓ Classic or digital television receiver[168]

Distribution of television

- ✓ Terrestrial free- to -air
- ✓ Satellite link
- ✓ Cables
- ✓ Internet

Principles of Analog Television Reception

- ✓ Reception of radio waves broadcast by the receiving antenna
- ✓ Channeling of the signals to the tuner
- ✓ Selection of wanted signals
- ✓ Demodulation of the modulated signal
- ✓ Sending of the electrical signal corresponding to the images on the screen
- ✓ Sending of the electrical signal corresponding to the sound to the speaker for conversion

Frequency band of an analog television channel (channel)

- ✓ 6 MHz for a channel or a television channel in America
- ✓ 8 MHz for a channel or a television channel in Europe
  - o *Frequencies corresponding to each television channel*

Use of the frequency band of a television channel

- ✓ Sound
- ✓ Image
- ✓ Luminance
- ✓ Chrominance

Frequency bands of analogue television

3 frequency bands

- ✓ 2 bands in the VHF (Very High Frequencies)

167 Guide pratique http://www.mediateur-telecom.fr/ressources/media/files/Guide_pratique_chapitre03.pdf

168 Numérique et multimédia - Digital Wallonia http://www.awt.be/web/img/index.aspx?page=img,fr,tel,010,010

- Low part in the VHF band: 54 - 88 MHZ
- High part in the VHF band: 174 - 216 MHz

✓ 1 band in the UHF (Ultra High Frequencies)

- Range: 470 - 890 MHz

Table of channels and associated frequencies

Band	Channel	Frequency (MHz)	Band	Channel	Frequency (MHz)
VHF	02	54 - 60	UHF	43	644 - 650
VHF	03	60 - 66	UHF	44	650 - 656
VHF	04	66 -72	UHF	45	656 - 662
VHF	05	76 - 82	UHF	46	662 – 668
VHF	06	82 - 88	UHF	47	668 – 674
VHF	07	174 - 180	UHF	48	674 – 680
VHF	08	180 - 186	UHF	49	680 – 686
VHF	09	186 - 192	UHF	50	686 – 692
VHF	10	192 - 198	UHF	51	692 – 698
VHF	11	198 - 204	UHF	52	698 – 704
VHF	12	204 - 210	UHF	53	704 – 710
VHF	13	210 - 216	UHF	54	710 - 716
UHF	14	470 - 476	UHF	55	716 - 722
UHF	15	476 - 482	UHF	56	722 – 728
UHF	16	482 - 488	UHF	57	728 – 734
UHF	17	488 - 494	UHF	58	734 – 740
UHF	18	494 - 500	UHF	59	740 – 746
UHF	19	500 - 506	UHF	60	746 - 752
UHF	20	506 - 512	UHF	61	752 – 758
UHF	21	512 -518	UHF	62	758 – 764
UHF	22	518 - 524	UHF	63	764 – 770
UHF	23	524 -530	UHF	64	770 - 776
UHF	24	530 - 536	UHF	65	776 – 782
UHF	25	536 -542	UHF	66	782 – 788
UHF	26	542 -548	UHF	67	788 – 794
UHF	27	548 - 554	UHF	68	794 – 800
UHF	28	554 - 560	UHF	69	800 – 806
UHF	29	560 - 566	UHF	70	806 – 812

UHF	30	566 -572	UHF	71	812 – 818
UHF	31	572 - 578	UHF	72	818 – 824
UHF	32	578 - 584	UHF	73	824 – 830
UHF	33	584 - 590	UHF	74	830 – 836
UHF	34	590 - 596	UHF	75	836 – 842
UHF	35	596 - 602	UHF	76	842 – 848
UHF	36	602 - 608	UHF	77	848 – 854
UHF	37	608 - 614	UHF	78	854 – 860
UHF	38	614 - 620	UHF	79	860 – 866
UHF	39	620 - 626	UHF	80	866 – 872
UHF	40	626 - 632	UHF	81	872 – 878
UHF	41	632 - 638	UHF	82	878 – 884
UHF	42	638 - 644	UHF	83	884 – 890

DIGITAL TELEVISION

- ✓ New TV system based on digital technologies
- ✓ New technologies for the processing and transmission of images, sound and data
- ✓ Digital television: television production, compression of images, multiplexing of signals, broadcasting of signals

Types of digital television

4 types

1.- Digital Terrestrial Television

Broadcasting of digital television programs by transmitters installed on the ground[169]

2.- Digital Cable Television

Broadcasting of digital television programs via a cable network

3.- Digital satellite television

Broadcasting of digital television programs by satellites of telecommunications

4.- Mobile digital television

Broadcasting of digital terrestrial television programs accessible on mobile terminals (smartphones, tablets, PDAs)

Principles of Digital Terrestrial Television (DTT)

- ✓ Digitization of video, audio and data signals
- ✓ Multiplexing of digitized signals (transforming digitized signals into a single stream)
- ✓ Modulation
- ✓ Broadcasting

Advantages of digital television

- ✓ Reception of more programs
- ✓ More varied television offers
- ✓ Quality of services: Best pictures and sounds, interactive services, etc.
- ✓ Digital equipment: More choices of digital TV receivers and decoders
- ✓ Digital platform: availability of a variety of services such as TV education, government online, telemedicine, teleshopping, etc.
- ✓ Liberated frequency for the deployment of 4G

169 Télévision numérique terrestre http://dlCTsionnaire.reverso.net/francais-definition/television-numerique-terrestre

services (new services, digital dividend)

- ✓ Lighter infrastructure for operators (new players in the digital landscape)
- ✓ More players on the audiovisual market
- ✓ Switch from standard definition (SD) to high definition (HD)
- ✓ Possibility to save content (shows, movies, etc.) to a hard drive or DVD
- ✓ Possibility to view multiple channels at the same time on the screen
- ✓ Integration of communication on the television: telephone calls, SMS, e-mail, Internet access, telebanking, online games, etc.
- ✓ Easy integration of interactive television
- ✓ Better frequencies management

Reception Types of DTT (Digital Terrestrial Television)

- ✓ Fixed reception with roof antenna and adapter
- ✓ Portable reception (reception of digital programs by an indoor antenna installed on the television, or even integrated
- ✓ Mobile reception

Reception of DTT via an analog receiver

- ✓ Antenna, Dish, Cable
- ✓ Decoder
- ✓ Display images

Reception of DTT via a Digital Receiver

- ✓ Antenna, Satellite, Cable
- ✓ Digital receiver
- ✓ Display images

Reception terminals for digital television

Several types of receiving terminals

- ✓ Analog TV and a digital adapter
- ✓ Built-in digital TV
- ✓ Computer equipped with a PC-TV tuner card
- ✓ Smart devices (smart GSM) capable of receiving mobile TV video[170]
- ✓ Flat screens installed in buses, trains and planes
- ✓ Game consoles

Architecture of the DTT network

- ✓ Programs (Editing content and programs)
- ✓ Compression (MPEG - 2/4)
- ✓ Multiplexing
- ✓ Transport
- ✓ Broadcasting (transmitters)

Components and roles

- ✓ Channel (TV station): Production of TV programs
- ✓ Transport: Routing of TV signals from the point of production of programs to the point of broadcasting of television signals
  - o Links: Studio - Multiplex
  - o Links: Mobile (live broadcast) - Studio
  - o Links: Multiplex - transmitter
- ✓ Multiplex: Grouping of several television programs to form a composite signal
- ✓ Broadcasting: Multiplexed signal radiation from a TV transmitter

Formats of digital television

2 options for resolution

Standard Definition (SD or SDTV)

- ✓ Format used for video production/reception with a picture quality level equivalent to conventional analog TV (720x576 or 640x480)

High Definition (HD or HDTV)

- ✓ Format used for video production/reception with a higher quality level of images, close to that of natural images
- ✓ Minimum format of high definition: 1080x720 (720 lines with 1080 pixels per line), and maximum HD format up to 4046x2048 (Digital Cinema 4k)

---

170 La télévision numérique - Agence Wallonne des Télécommunications http://www.awt.be/web/img/index.aspx?page=img,fr,tel,020,005.

Digital Television Standards

- ✓ ATSC: Advanced Television Systems Committee (United States of America)
- ✓ DVB: Digital Video Broadcasting (Europe)
- ✓ ISDBT - T: Terrestrial Integrated Digital Broadcasting Services (Japan)
- ✓ DTMB: Digital Terrestrial Multimedia Broadcasting (China)
- ✓ SBTVD-T: Sistema Brasileiro of Televisão Digital Terrestrial (Brazil)

Variants of DVB (Digital Video Broadcasting)

- ✓ DVB-T: for digital terrestrial transmissions (DTT)
- ✓ DVB-C: for cable transmissions
- ✓ DVB-S and DVB-S2: for digital satellite transmissions
- ✓ DVB-H: DVB-T version adapted for mobile transmissions

Players of Digital Terrestrial Television (DTT)

Several players

Service or program publishers

- ✓ Production, co-production or purchase of audiovisual programs for public distribution
- ✓ Production of audiovisual programs
- ✓ Compliance with the control rules by the regulatory authority
- ✓ Prior authorization of the regulatory authority prior to selecting a multiplex operator

Multiplex operators

- ✓ Assembly of signals (multiplexing) from the service editors for their transport to the broadcaster (the broadcasting operator)
- ✓ Compliance with the rules of the regulatory authority

Technical broadcasters or broadcasting operators

- ✓ Broadcasting of radio signals to the public (over-the-air, satellite, cable or cable television programs)
- ✓ Compliance with rules established by the telecommunication's regulatory authority

Commercial Distributors

- ✓ Companies responsible for distributing services to the public[171]

Transition from analog Television to digital television

Adoption process of digital television

Abandonment of analog television systems

Stake-holders of the transition to digital television

- ✓ Government (Major decisions concerning the choice of the standard of DTT, possible subsidy of the decoders)
- ✓ Regulator (Institutional Guarantor for the Development of the Audiovisual Sector)
- ✓ Television operators (use of digital technologies for the provision of audiovisual services)
- ✓ Telephony operators (use of the digital dividend resulting from the transition for the provision of 4G services)
- ✓ Equipment Vendors (Provision of Digital Television Receiving Terminals)
- ✓ Professionals (Training of human resources for the use of digital systems)
- ✓ Consumers (Decoders and New Receivers to Purchase for digital TV Access)

Cable television (Teledistribution)

- ✓ Cable TV (CATV: *Community Access Television* or *Community Antenna Television)*
- ✓ Telecommunication intended for the distribution of visual or sound programs to certain users by cable networks (coaxial cables or optical fibers)
- ✓ Mode of distribution of television programs via a cable network

Main elements of a cable system

- ✓ Head -end
- ✓ Cable television network
- ✓ Transmission cables
- ✓ Decoder (set up box)

171 La télévision numérique terrestre - David ARNOULT http://morin80s.free.fr/TNT/MemoireProbatoireTNTDavidARNOULT.pdf

- ✓ Terminal (TV or specific receiver)

Principle of Cable Television

- ✓ Capture of still and moving images by the camera
- ✓ Conversion of images into an electrical signal by the camera
- ✓ Various signal processings
- ✓ Transmission of the electrical signal via a cable
- ✓ Reception of the electrical signal by a television receiver
- ✓ Various treatments
- ✓ Conversion of the electrical signal into images, and display on the screen

Principle of operation

- ✓ Distribution by the head-end of television programs
- ✓ Reception by each subscriber via a cable of the same programs

Cable TV reception

- ✓ Subscription to the cable television service
- ✓ Classic television receiver

Frequency bands reserved for Cable Television

- ✓ Frequency spectrum allocated (50 -550 MHz: 80 channels)
- ✓ 6 MHz bandwidth usage by each program

IP Television (IP TV)

- ✓ Communication protocol used for the transmission and reception of television services via an Internet connection
- ✓ Broadcasting of TV programs performed by the Internet Protocol (IP)
- ✓ Television programs broadcast by Internet boxes (Catch-up TV and on-demand TV) Audiovisual content delivered via an Internet connection

Principle of operation

- ✓ Coding of video streams as IP packets
- ✓ Broadcasting of video data across IP networks
- ✓ Conversion of video data to video signal for terminal screens

Types of IP TV

- ✓ Real time broadcasting
- ✓ Video on demand (access to videos stored on servers) / access to a video catalog
- ✓ Time-shifting videos: Catch up tv, TV Start -over

IP Television Services

- ✓ Classic channels (digital TV services)
- ✓ Video on demand (VoD)
- ✓ Near video on demand (nVoD)
- ✓ Time - shifting
- ✓ Catch Up TV

Components of an IP TV system

- ✓ Head-end (suitable for IP transmission)
- ✓ Core network (IP or MPLS network)
- ✓ Access network
- ✓ User equipment

IPTV receiving terminals

- ✓ Computer (with installation of specific software)
- ✓ TV screen equipped with set-box
- ✓ Digital tablet
- ✓ Cellular telephone

Access to IP TV

1.- Free Access (broadcasting of live channels)

- ✓ Internet connection and a terminal compatible with Internet (computer)

2.- Paid access (video on demand)

- ✓ Set top box
- ✓ Internet connection

Means of Reception of IPTV

- ✓ Appropriate terminal
- ✓ Internet connection

Advantages of IP Television

- ✓ Interactive television
- ✓ Less bandwidth occupied
- ✓ Protection of signals against noise and ghost signals

- ✓ Possibility of storage of videos on a server for later viewing
- ✓ Lower cost for deployment
- ✓ Bilateral communication (pauses, backtracking, backup, etc.)

Disadvantages of IPTV

- ✓ Sensitivity to packet loss
- ✓ Delay due to the mode of transmission
- ✓ Internet connection speed
- ✓ Latency (satellite transmission)

Internet TV

Web TV, TV online, Net TV

- ✓ Broadcast and reception by a web interface of video signals[172]
- ✓ Method of distributing and transmitting multimedia content via the Internet[173]
- ✓ Video content broadcast over the Internet

Technology used

- ✓ Streaming for content broadcasting

Principles of Operation

- ✓ Use of IP infrastructures (DSL, Wi Fi, 3G)
- ✓ Distribution model
  - o Live streaming
  - o Video on demand

Reception of Web TV

- ✓ Computer, tablet and mobile phone
- ✓ Internet connection
- ✓ Web browser
- ✓ Media player

---

172 Web TV
h ttps://fr.wikipedia.org/wiki/Web_TV

173 Les réseaux de distribution du contenu audiovisuel http://www.awt.be/web/img/index.aspx?page=img,fr,tel,020,010

Chapter 7

COMPUTING AND INTERNET

COMPUTING OR INFORMATICS

Functionality and operations based on appropriate machines and programs

- ✓ *Computing:* Science of automatic information processing by programs installed on computers
- ✓ *Computing*: Scientific, technical and industrial field of activity relating to the automatic processing of information
- ✓ *Computing*: Science of rational processing, especially by automatic machines, of information considered as the support of human knowledge and communications in the technical, economic and social fields[174]
- ✓ *Computing*: Set of tasks and applications automation technologies
- ✓ *Computing*: Set of techniques for collecting, sorting, storing, transmitting and using automatically processed information using programs implemented on computers[175]

Basics of Computing

- ✓ Hardware
  - o Device developed to run using software
  - o Computers, Servers, Robots, automaton, Embedded Systems, etc.
- ✓ Software
  - o Logical programs installed on computers, tablets, smartphones for their operations
  - o Examples: Operating Software, Application Software and Development Software

Computer

- ✓ *Computer*: Automatic information processing machine, obeying programs formed by sequences of arithmetic and logical operations[176]
- ✓ *Computer*: Programmable electronic device for storing, retrieving and processing all kinds of information (sound, speech, image, videos, texts, data)
- ✓ *Computer*: Terminal necessary for the user to access computer services (personal use, access to local networks or Internet)
- ✓ *Computer*: Electronic machine composed of several parts interconnected by wires
- ✓ *Computer*: Machine originally intended to make huge numerical calculations

Components of a computer

- ✓ Input devices (keyboards, mice, microphones, cameras, etc.)
- ✓ Processor
- ✓ Memory
- ✓ Output devices (screen, printer, speaker, projector)

Hardware parts of a computer

Main components of a computer

- ✓ Computer Hardware: Set of spare parts for information processing
- ✓ Power supply
- ✓ Motherboard
- ✓ Processor
- ✓ Fan
- ✓ RAM memory
- ✓ Hard disk
- ✓ CD / DVD player and burner
- ✓ Graphic card

Criteria of choice of computer

3 basic criteria for all types of computer

- ✓ Processor speed (expressed in GHz)
- ✓ Memory capacity (expressed in GB)
- ✓ Storage capacity (expressed in GB or TB)

Software

- ✓ *Software*: All programs, processes and rules for data processing
- ✓ *Software*: Suites of instructions describing in detail the algorithms of information processing op-

174 Futura Tech – http://www.futura-sciences.com/tech/definitions/informatique-informatique-553/

175 Informatique – http://encyclopedie_universelle.fracademic.com/10309/INFORMATIQUE

176 Larousse – http://www.larousse.fr/dictionnaires/francais/ordinateur/56358

erations

- ✓ *Software:* Programs required for data entry, processing, output, storage and control of information systems activities
- ✓ *Software*: Set of programs cooperating to perform a particular task
- ✓ *Software*: Computer Intelligence
- ✓ *Software*: Essential tool for operating the computer

Types of software

- ✓ Operating system
- ✓ Application software
- ✓ Development Software
- ✓ Communication Software (Network Operations and Management)

Operating systems

- ✓ Software designed to operate the computer, server, cellular telephone

Main operating systems

- o Windows 7
- o Windows 8
- o Windows 10
- o Linux
- o Mac OS X
- o IOS (Cellular telephone)
- o Android (Cellular telephone)

Functions of an operating system

Control of the resources of a computer system

- ✓ Assignment of needed hardware to Programs or Software Packages
- ✓ Programming of programs or software packages to be executed by the processor
- ✓ Allocation of memory for each program
- ✓ Assignment of devices needed for entry and output
- ✓ Management of data and program files stored in a secondary storage
- ✓ Maintaining of file directories
- ✓ Provision of access to file data
- ✓ Interaction with users

Functions of an application software

- ✓ Achievement of specific tasks
  - o Word Processing (WORD)
  - o Calculation (EXCEL)
  - o Database (ACCESS)
  - o Presentation (POWER POINT)
  - o Web browsing (Internet Explorer, Google chrome, Safari)
  - o Multimedia playback (VLC media player, 5KPlayer)

Functions of a development software

- ✓ Software or computer program development
  - o C
  - o C ++
  - o Java
  - o Delphi
  - o Visual Basic
  - o Python

Computer language

- ✓ Computer alphabet: 2 "letters" or 2 elements called Bit
- ✓ Binary language: Only language understood and used by the computer
- ✓ *Bit*: contraction of ***binary digit***
- ✓ *Bit*: smallest unit of information manipulated by a digital machine, represented by a "1" or a "0"
- ✓ *Bit*: Form adopted by the information to be processed and transmitted in digital systems (Bit conversion (coding) of information of all kinds)
- ✓ *Bit*: Basic element in the operation of digital telecommunications systems
- ✓ *Bit*: Form taken by all information (speech, music, text, still images, videos and data) transmitted

through a digital telecommunications system

- ✓ *Bit*: "Currency of exchange" in the universe of Computing and Telecommunications
- ✓ *Binary sequence*: sequence of bits, that is to say, sequence consisting of "" 1 "and" 0 "(for example: 1111 or 0000 or 0011000111100000 or 11111110000000
- ✓ *Binary sequence*: Mean to represent by bit a letter, a number, a word, a sentence
- ✓ *Binary word*: Word formed of a set of bits
- ✓ Letter "A": 01000001 (for example on an 8-bit coding)
- ✓ *Digital file*: File formed of a set of bits

Relations between Bit and Byte

- ✓ 1 Byte = 8 bits
- ✓ 10 Bytes = 80 bits
- ✓ 800 bits = 100 Bytes

Multiples of Bit and Byte

Important computer rules

- ✓ 1 Kilobit = 1024 bits (1024 = 2^{10}, closer to 1000 = Kilo)
- ✓ 1Mega bit = 1024 Kb = 1048 576 bits
- ✓ 1Giga bit = 1024 Megabits = 1048 576 K bits = 1073,741,824 bits
- ✓ 1 Terabit = 1024 Giga bits = 1048576 Megabits = 1073 741 824 Kb = 1099 511 627 776 bits
- ✓ 1 Pétabit (1 Pb) = 1024 Tb = 1 125 899 906 842 624 bits
- ✓ 1 Exabit (1 Eb) = 1024 Pb
- ✓ 1 Zettabit (1Zb) = 1024 Eb
- ✓ 1 Yottabit (1 Yb) = 1024 Zb

Similarly, for the Byte (B)

- ✓ 1 Kilo byte (KB) = 1024 Bytes
- ✓ 1 Megabyte (MB) = 1024 KB = 1048,576 Bytes
- ✓ 1 Gigabyte (GB) = 1024 MB = 1048 576 KB = 1073,741,824 Bytes =
- ✓ 1 Terabyte (TB) = 1024 GB = 1048 576 MB = 1073 741 824 KB = 1099 511 627 776 Bytes
- ✓ 1 Petabyte (PB) = 1024 TB = 1048 576 GB = 1 073 741 824 MB = 1,125,899,906,842,624 bytes
- ✓ 1 Exabyte (1 EB) = 1024 PB = 1, 152, 921, 504, 606, 846, 976 bytes
- ✓ 1 Zettabyte (1 ZB) = 1024 EB = 1, 180, 591, 620, 717, 411, 303, 424 bytes
- ✓ 1 Yottabyte (1 YB) = 1024 ZB =1, 208, 925, 819, 614, 629, 174, 706, 176 bytes
- ✓ 1 Brontobyte (1 BB) = 1024 YB
- ✓ 1 Geopbyte) (1 GPB) = 1024 BB

Size of digital information

- ✓ Size = volume of information = weight in bits or bytes of information
- ✓ Bit and Byte = Units of information stored or transmitted
- ✓ Data storage medium: MB, GB and TB
- ✓ Little message of a few words: 5KB
- ✓ 2-minute telephone conversation on WhatsApp: 1 MB
- ✓ E-mail (text only, without attachment): 5 KB to 25 KB
- ✓ Email with attached photos: 400 KB Size of a compressed photo: 500 KB (approximately)
- ✓ Uncompressed photo: 2 to 5 MB
- ✓ Web page: 150 KB to 1.5 MB
- ✓ One minute of music in streaming mode: 500KB
- ✓ Short video = 2 MB
- ✓ One minute of streaming video: 3MB to 7MB
- ✓ Downloading of games and songs applications: 4MB to 10 MB
- ✓ Short movie: 5 GB

Data storage medium

Main storage media

- ✓ Floppy disks (1.44 MB)
- ✓ Compact disk (CD: 700 MB)

- ✓ Digital versatile disc (DVD: 4.7 GB or 9.4 GB dual layer)
- ✓ USB Key (from 1GB to 128GB)
- ✓ Memory cards for digital devices (from 2 GB to 128 GB)
- ✓ Internal computer hard disks (hundreds of GB to TB)
- ✓ External computer hard drives (hundreds of GB to TB)

Database

- ✓ Collection of data organized to be easily accessible, administered and updated[177]
- ✓ Database: Stored, organized and structured data for software queries
- ✓ Database: Tool designed to store and retrieve information

Computer Networks

- ✓ Set of interconnected computers and terminals to exchange digital information
- ✓ Means of communication, devices, and software necessary to connect computer systems and/or devices
- ✓ Set of computer resources (**materials** and **software**) implemented to ensure the sharing of information between computers, workstations and computer terminals
- ✓ Transmission media, devices and software necessary to connect two or more computer systems and/or device

Goals of computer networks

- ✓ Sharing information between a group of remote users
- ✓ Sharing hardware (printers), programs and database
- ✓ Facilitating teamwork, innovative ideas, and new business strategies
- ✓ Etc.

Types of computer networks

- ✓ *Local Network*: Link between computers and peripherals (eg printer) located close to each other, for example in the same building)
- ✓ *Network local* : Network widely used in companies and institutions
- ✓ *Local network size*: up to 100 computers
- ✓ *Data transfer speed*: 10 Mbps to 1Gbps
- ✓ *Metropolitan Network*: Set of Local Area Networks
- ✓ *Metropolitan Network*: Interconnection of several geographically close local networks
- ✓ *Metropolitan Network*: Range of a few tens of kilometers
- ✓ Wide area network: Interconnection of several local networks in a large geographical area or several metropolitan networks in the same country or in the world
- ✓ Example of WAN: Internet

Computer Network Software

- ✓ Communication software
- ✓ Network operating systems
- ✓ Network Management Software

Functions of communication softwares

- ✓ Error check
- ✓ Formatting of messages
- ✓ Communications Accounting
- ✓ Security and confidentiality of data
- ✓ Translation Capabilities for Networks

Functions of network operating systems

- ✓ Control of computer systems and devices
- ✓ Means of communication between computer systems and devices

Functions of network management software

- ✓ Control of the use of personal computers from a desktop (networked) and hardware sharing by a manager
- ✓ Virus detection
- ✓ Software license compliance

177 LemagIT – http://www.lemagit.fr/definition/Base-de-donnees

Computer Network Infrastructure

- ✓ Hardware devices
  - o Connectivity
  - o Routing
  - o Means of commutation
  - o Security and access control
- ✓ Software components
  - o Operating system
  - o Other specific software

Network hardware

- ✓ Servers (network resources management)
  - o File server, web server, print server, email server
- ✓ Routers (multiple network connections)
- ✓ Switches (packet forwarding to destination port only)
- ✓ Hubs (device capable of collecting and distributing data communications in a star network)

Computer servers

- ✓ *Computer server*: computer with huge processing capabilities
- ✓ *Computer server*: tool designed to answer users' questions 24 hours a day
- ✓ *Computer server*: computer designed to provide information and software to other computers over a network (A search by a user from his personal computer connected to the Internet)
- ✓ *Computer server*: Specific computer sharing resources with other computers called clients
- ✓ *Computer server* : essential tool for all online services

Functions of a computer server

Supply of information and software to computers connected via a network

- ✓ Storage of files and data
- ✓ Search for files and data
- ✓ Transmission of files and data
- ✓ Sharing of information

Types of computer servers

- ✓ File server
- ✓ Print Server
- ✓ Email server
- ✓ Web server
- ✓ Backup server
- ✓ Application server
- ✓ Streaming media server

Services of a computer server

Most common services of a server

- ✓ Access to information available on the world wide web (www)
- ✓ Email
- ✓ Remote consultation of a data bank
- ✓ File and printer sharing
- ✓ Remote diagnosis
- ✓ Database storage
- ✓ Management of authentication and access control
- ✓ Online games
- ✓ Supply of application software

Fields of use of computer servers

- ✓ Computer Networks
- ✓ Internet
- ✓ Telecommunications Operations
- ✓ Data Processing Center
- ✓ Etc.

Operating requirements of computer server

- ✓ Operating system
- ✓ Database
- ✓ Web applications

Data communication model

- ✓ Input information (source)
- ✓ Input data or bit stream (transmitter)

- ✓ Transmitted signal (transmission system)
- ✓ Received signal
- ✓ Receipt of stream (receiver)
- ✓ Output data
- ✓ Output Information (Destination)

Data transmission

Data transmission: digital transmission or digital communications

- ✓ *Data transmission*: Transfer of bitstreams or digitized analog signals via point-to-point or point-to-multipoint communication channels
- ✓ Point-to-point, point-to-multipoint and multi-point-to-multipoint data transmission
- ✓ Sending and reception of digital data

Principles of data transmission

- ✓ Acquisition of data
    - o Entry of data using keyboard and screen
    - o Digitization of analog data
- ✓ Data processing
    - o Data conversion
    - o Data adaptation
    - o Data control
- ✓ Transmission of data
    - o Sending of data via a transmission medium (cables, wireless links, optical fiber, data bus)
    - o Data transfer (guided or unguided propagation of signals)
    - o Transmission provided by the DCE (data circuit terminating equipment or data communication equipment)
- ✓ Data analysis
    - o Data reception and decoding
    - o Display of data on a computer or a terminal with screen

Data transmission infrastructure

- ✓ Computer terminals (Computers or other terminals)
- ✓ DTE (Data Terminal Equipment)
- ✓ Computer, screen terminal, printer
    - o Source/ collector of data
    - o Communication controller
- ✓ Digital Interfaces (exchange between DTE and DCE)
- ✓ DTE (Data Circuit Termination Equipment = Modem)

Modem, Multiplexer

- ✓ Transmission medium
- ✓ Data circuit
- ✓ Data link

End components of a data transmission

- ✓ Computer connected to the network (eg Internet)
- ✓ Host computer permanently connected to the Internet (web server)
- ✓ Device (printer)

Bit rate (binary speed or throughput)

- ✓ *Bit rate*: number of bits transmitted per second on a transmission medium
- ✓ *Bit rate*: Amount of information transmitted via a communication channel during a given time interval
- ✓ *Bit rate*: speed of information transmission
- ✓ *Bit rate*: connection speed between the network and the user

Units of the bit rate (Data Transmission Speed)

- ✓ Bit rate unit: bit per second or bit /second or bps or b/s
- ✓ Byte per second, Byte /second
- ✓ Multiple Bit Rate: Kilobits per second (1024 bits / second)
- ✓ Megabits per second (1024 Kbits / second)
- ✓ Gigabits per second (1024 Mbps)
- ✓ Terabits per second (1024Gb /second)

Some examples of digital bit rates

- ✓ Speech signal: 64 Kbps (Digital telephony line rate = 64 Kilobits / second (transmission of 64,000 bits per second)
- ✓ Bitrate for Color Video Conferencing: 100 Mbps
- ✓ Bitrate for Color Television: 204 Mbps[178]
- ✓ Internet connection at 1 Gb /s: Connection establishing a link transmitting more than a billion bits per second
- ✓ 6 seconds to download a 75-byte file with a speed of 100 bits per second
- ✓ 75 minutes to download a movie of 2 gigabytes of data (2 GB) with a speed of 4 megabits per second
- ✓ Downloading of 10 MB file (10 MB):
    - o 25 minutes for a bit rate of 56kb/s (low bit rate)
    - o 2 minutes 30 seconds for a speed of 512kb/s (broadband)
    - o 8 seconds for a bit rate of 10 Mb/s (broadband)
    - o 2 seconds for a speed of 50 Mb/s (very high speed)
    - o 1 second for a bit rate of 100 Mb/s (very high speed)

Telecommunications and Computers

Fusion: Telecommunications and Computers

- ✓ Contribution of Computing to telecommunications
- ✓ Contribution of telecommunications to Computing
- ✓ Bridging of two distinct disciplines

Uses of Computing in Telecommunications

Hardware and software used in the automation of telecommunications operations

- ✓ Computer: terminal operated to access and use telecommunications services
- ✓ Use of software in the automatic processing of information (storage, processing, etc.)
    - o Example: Registration of customers' accounts and transactions data management in databases

Contribution of intelligence and automation to telecommunications

- ✓ Use of software intelligence in the operations of telecommunications
- ✓ Control software: software to verify the functionality of systems
    - o Example: electronic recharge of user accounts, generation of balance of accounts via USSD, management of subscriber accounts by computer servers
    - o Example: Processing and answers to queries of millions of telephone subscribers by computer servers

Uses of telecommunications in computing

Telecommunication infrastructures deployed for transporting data from computers

- ✓ Establishment of links between computers
- ✓ Transport of signals generated by computers
- ✓ Networking using equipment of telecommunications and transmission media
- ✓ Example: Communication between two remote computers supported by a transmission medium

Contribution of networking to Computing

- ✓ Development of networked computing
- ✓ Deployment of local networks
- ✓ Deployment of metropolitan networks.
- ✓ Deployment of wide area networks
- ✓ Example: Transport (transmission) throughout the world of the signals of the Internet

Tele informatics

- ✓ Association of Telecommunications and Computer Technologies for the remote processing of information
- ✓ Set of techniques implementing aspects of telecommunication in the service of computing[179]
- ✓ Telecommunications + Computing = remote access and data processing

178 Modulations analogiques et numériques
Polytech A5 Ibats D. Vivet damien.vivet@univ-orleans.fr

179 Introduction à la téléinformatique – http://dept-info.labri.fr/~felix/Annee2005-06/AS/Teleinformatique/Livret%20I.pdf

- ✓ Telecommunications + Computing = Decentralization of information processing
- ✓ Telecommunications + Computing = Execution of work by separate, but interconnected computers

Applications of Tele informatics

- ✓ File transfer between computers
- ✓ Cooperative processing between 2 applications
- ✓ Distributed database
- ✓ Sharing of resources in a network of microcomputers[180]

Technological convergence or digital convergence

- ✓ Integrations between computer technologies and telecommunications
- ✓ Bridging of different sectors, mainly telecommunications, audiovisual and computing
- ✓ Merger of three sectors (Audiovisual, Telecommunications and computing) operating once separately

Fields of "Telecommunications"

- ✓ Telegraphy
- ✓ Telex
- ✓ Telephony
- ✓ Fax
- ✓ Etc.

Fields of Broadcasting

- ✓ Broadcasting of sound
- ✓ Broadcasting of images

Fields of Computing

- ✓ Automatic processing of digitized information

Basis of technological convergence

- ✓ Digitization of all types of information
  - o Digitized information suitable for storage, processing and transmission
- ✓ Digitization of processing and transmission systems
- ✓ Integration of audiovisual, telecommunications and computing functionalities in systems and terminals

Advantages of technological convergence

- ✓ Provision by a single network services formerly provided by several different networks (telephone network, television channel, Internet)
- ✓ Access via a single terminal to several services formerly accessible on separate terminals (telephone, television, computer)

Applications of technological convergence

- ✓ Mobile telephones connection to Wi-fi and hot spots for data services and calls
- ✓ Access to the Internet from a 3G or 4G modem inserted into the computer
- ✓ Use of the cellular telephone for internet access and television
- ✓ Availability of value-added services (use of prepaid telephone card, request for telephone credit balance)

Levels of technological convergence

Different levels of convergence

Convergence of networks

- ✓ Telephone telecommunications networks
- ✓ Data transmission networks between computer terminals
- ✓ Broadcasting networks for audiovisual services (radio, television, etc.)

Convergence of services

- ✓ Telephone services
- ✓ Computing services
- ✓ Multimedia Services

Convergence of transmission channels

- ✓ Transfer of all types of communication (radio, TV, Telephony, Internet) via one channel

Convergence of information media

- ✓ Recording of all types of information (computer data, images, videos) on CDs and DVDs

180 Introduction à la téléinformatique – http://dept-info.labri.fr/~felix/Annee2005-06/AS/Teleinformatique/Livret%20I.pdf

Convergence of applications and commercial offers

- ✓ Multifunction of integrated communications applications
- ✓ Offering of packages of services by telecommunications operators (eg triple play)

Convergence of terminals

- ✓ Other functions of a personal computer
  - o Telephone terminal
  - o Radio receiver
  - o TV receiver
- ✓ Other Functions of a mobile telephone terminal
  - o Camera
  - o Camcorder
  - o Electronic agenda
  - o Multimedia player (music, photos, movies, documents)

Combination of features supported by technological convergence

- ✓ TV receiver
  - o Access to radio
  - o Access to internet
- ✓ **Computer**
  - o Access to the radio
  - o Access to television
  - o Access to internet
- ✓ **Smartphone**
  - o Access to internet
  - o Access to radio[181]
  - o Access to television

Advantages of technological convergence

- ✓ One terminal (access to telephony, television, sound and internet)
- ✓ One network (provision of services: telephony, television, Internet, etc.)

181 Convergence numérique
https://fr.wikipedia.org/wiki/Convergence_numerique

Players of technological convergence

- ✓ Cable, terrestrial and satellite broadcasters
- ✓ Publishers and editors of printed works, video and audio works
- ✓ Telephone Manufacturers and Telecommunications Operators
- ✓ Software Developers
- ✓ Internet Content Editors
- ✓ Service Providers
- ✓ Database developers[182]

INTERNET

Internet: network of networks

- ✓ *Internet*: network of millions of small private networks, each with the ability to operate independently or in harmony with all the other millions of networks connected to the gigantic network
- ✓ Network consisting of computers and other devices logically interconnected by a single address based on the Internet protocol
- ✓ *Internet*: Global Information Infrastructure
- ✓ Internet: Inter and networks = interconnection of computer networks by telecommunications links
- ✓ *Internet* : international network of computers communicating with each other using standard data exchange protocols
- ✓ *Internet* : communication tool using telephone wires, optical fibers, intercontinental cables and satellite communications to make services such as e-mail and the world wide web available to users
- ✓ *Internet* : set of connected academic, military, financial and scientific networks.

Origin of the Internet

- ✓ Development in 1969 of the first worlwide network "ARPANET" by ARPA agency (Advanced Research Project Agency) of the United States of America

182 Telecommunications Law
Ian Lloyd and David Mellor

- ✓ Independent "ARPANET" network of a nerve center to guarantee its resilience in case of a nuclear attack
- ✓ Improvement of the network between 1970 and 1980 by the introduction of protocols facilitating exchanges between servers
- ✓ Availability of e-mail and File Transfer Protocol services through the introduction of protocols
- ✓ Replacement of ARPANET by the Internet in 1990
- ✓ Introduction of hypertext links in 1992

Initial spirit of the Internet

- ✓ Sharing of resources
- ✓ Open Access
- ✓ No government involvement in the regulation of the Internet

Basics of Internet Operation

- ✓ Protocol (communication between computers)
- ✓ Backbone network (high-speed long-distance links)

Main protocols of the Internet

2 Main: TCP and IP

- ✓ TCP / IP (Transmission Control Protocol / Internet Protocol): 2 Protocols used by Internet

Functions of Transmission Control Protocol (TCP)

- ✓ Management of the movement of data between computers
- ✓ Establishment of connection between computers
- ✓ Sequencing of packet transfer
- ✓ Acknowledgment of receipt for packets sent

Functions of (IP) Address

- ✓ Unique identification of each computer or other terminals connected to the Internet
- ✓ Protocol necessary for any exchange via the Internet (communication between the connected machine and other connected computers)
- ✓ IP: Addressing and routing mechanism
- ✓ Transmission and reception of data
- ✓ Addressing of packets transmitted over the Internet
- ✓ Assembly and disassembly of packets during transmission
- ✓ Delivery of packages

Principles of operation of Internet

- ✓ Use by each computer and network on the Internet of the same protocols (rules and procedures) to control timing and data format
- ✓ *TCP / IP*: Group or suite of networking protocols used to connect computers on the Internet
- ✓ Exchange of data on the Internet between computers by TCP/IP (whatever the computer)
- ✓ TCP: Provision of transport functions
- ✓ Quantity of data sent = Quantity of data received through TCP
- ✓ IP: unique address for each computer connected to the Internet
- ✓ Operation of the Internet based on packet switching
- ✓ Decomposition of the message into a data packet
- ✓ End-to-end data transfer as packets
- ✓ Packet = set of bytes with a specific size
- ✓ Address of the sender and the addresse of the recipient attached to each packet during transmission
- ✓ A serial number for each packet (ex: 1, 2, 3 ....)
- ✓ Independent transmission of a packet from other packets (possible arrival of packet 7 before 2, or that of 14 before 9) through the transmission network (backbone)
- ✓ Different routes taken by packets of a message
- ✓ Random reception of packets
- ✓ Gathering of packets in order of their numbers to the destination to restore the initial message
- ✓ Restitution of the original message

Value chain of the Internet

5 basic elements

- ✓ **Content rights**
  - o Videos, music, games
- ✓ **Online services**
  - o Online retail
  - o E-commerce
  - o Music
  - o Video
  - o Online travel
  - o Publication
  - o Media
  - o Games
  - o Online gambling
  - o Social networks
  - o Communication Services
  - o Search (Google, Baidu, Yahoo)
  - o Information and reference
  - o Cloud services (cloud computing)
- ✓ **Innovative technologies and services**
  - o Design and hosting (web hosting)
    - Examples: Go Daddy, Ipower
  - o Payment platforms
    - Examples: Alipay, Mastercard, Visa, Paypal,
  - o M2M payment
    - SIM management and M2M payment
    - Examples: Bosch, cumulocity
  - o Online Advertising
  - o Internet Analytics
  - o Bandwidth Management and Content Delivery
- ✓ **Connectivity**
  - o Fixed access (VPN, Wi-Fi, fixed operators)
  - o Mobile access (mobile operators)
  - o Satellite access (satellite operators: Iridium, Global star, Inmarsat)
- ✓ **User Interfaces**
  - o Personal computers
  - o Smart Phones
  - o Smart TVs
  - o Set -up box and digital receivers
  - o Digital tablets
  - o Consoles
  - o Other smart terminal devices
  - o Other materials
  - o Systems and software (operating systems, Apple store, security and software)[183]

Operating requirements of the Internet

- ✓ Server (hosting of information)
- ✓ IP Address (Assignment of an IP Address to the Server)
- ✓ Domain Name Server (direction to the IP address of the server)
- ✓ Internet Service Providers (setting up of a gateway for connection with the rest of the Internet)
- ✓ Browser or navigator (software for navigation)
- ✓ User (creation of information or access to information available via a browser)

Types of Internet infrastructure

3 types of infrastructure

1.- Physical infrastructure: Equipment

Hierarchy of interconnected networks

- ✓ Local networks
- ✓ Departmental networks
- ✓ Campus or corporate networks
- ✓ Wide Area networks

Networking and Interconnection Devices

183 The Internet value chain https://www.gsma.com/publicpolicy/wp-content/uploads/2016/05/GSMA_The-internet-Value-Chain_WEB.pdf

- ✓ Router: element allowing the referral of the data
- ✓ Hub: Equipment connecting several elements of a network
- ✓ Gateway: element allowing connection of networks of different nature

2.- Access infrastructure: Links between the user and the Internet

Data transmission media (multimedia signals)

- ✓ Electrical cables (electrical signals)
- ✓ Free space (electromagnetic waves)
- ✓ Optical fiber (light signal)

Types of links used

- ✓ Digital Subscriber Line (64kbps)
- ✓ ISDN line (128 kbps)
- ✓ E1 (2 Mbps)
- ✓ T1 (1.544 Mbps)

Transmission network

- ✓ Wired links (twisted pairs, coaxial cables)
- ✓ Optical Fiber
- ✓ Microwave links
- ✓ Telecommunications Satellites

3.- Communication Infrastructure: Protocols

Rules governing the operation of machines connected to the Internet (communication between machines)

- ✓ Definition of the format and order of messages exchanged between one or more communicating entities
- ✓ Definition of the actions to be taken for the transmission and / or reception of a message
- ✓ Methods of data transmission
- ✓ Communication language between machines for data transmission

Main protocols used by the Internet

- ✓ Transmission Control Protocol / Internet Protocol (TCP / IP): Establishment of rules for the flow of information over the Internet
- ✓ Hypertext Transfer Protocol (http): tool used to visualize a website through a browser

Main Elements of Internet Infrastructure

- ✓ Backbone network
- ✓ Internet Exchange Point (IXP)
- ✓ Regional Networks
- ✓ Local internet access providers

Components of the Internet network

- ✓ Data Center
- ✓ Storage devices
- ✓ Telecommunications links
- ✓ Operating systems (software)
- ✓ Web servers, Internet storage
- ✓ Computer Networks
- ✓ Computer servers
- ✓ Storage devices (storage for accessibility)
- ✓ Switches
- ✓ Stakeholders of the Internet
- ✓ Content Providers
- ✓ Hosting Providers
- ✓ Access Providers
  - o Telecoms Operators and Internet Service Providers
- ✓ **ICANN**
  - o Management authority for so-called first-level extensions of domain names
  - o Management of the IP addressing system of the web
- ✓ **Registry (AFNIC, VERISIGN, EURID)**
  - o .com Management (VERISIGN)
  - o .fr Management (AFNIC)
- ✓ Registrars / GANDI, OVH
  - o Direct sales to end customers
  - o Update of DNS (Domain Name Server)

- o DNS: basis of correspondence between IP addresses and domain names

✓ Web Hosts

- o Commercial entity providing servers and networks for the provision of websites and web services
- o Role of registrar
- o Internet access
- o Transfer of data to consumers

✓ W3C (World Wide Web Consortium)

- o Standardization body responsible for defining and promoting World Wide Web technologies (HTML, CSS, XHTML, XML, PNG, etc.)

Internet ecosystem

Different elements

✓ Experts in Technology, Engineers, Architects, Designers and Organizations (Internet Engineering Task Force (IETF) and World Wide Web Consortium (W3C)

- o Role: Coordination and implementation of open standards

✓ International and local organizations ICANN (Internet Assigned Names and Domains), IANA (Internet Assigned Numbers Authority), RIR (Regional Internet Registries), and Registries and Registrars of Domain Names

- o Role: Management of worldwide addressing resources

✓ Operators, Engineers and Providers Network Operator and Internet Exchange Points (IXP)

- o Role: Provision of telecommunications network infrastructures and provision of domain names

✓ Users

- o Role: Use for Communication purpose, and service offering

✓ Educators, Multilateral Organizations, Educational Institutions and Government Agencies

- o Role: Knowledge Education and Capacity Building in the Development and Use of Internet Technologies

✓ Decision - makers

- o Role: Local and Global Policy Making and Internet Governance[184]

Internet management

✓ Management entrusted to organizations (UN / ITU, IAB, ISOC, ICANN IETF, W3 Consortium, IANA, ASO, NIR,)

✓ Agency Tasks: Standards Development, Domain Name Assignment and IP Addresses

✓ Internet Resource Management (Domain Names and Addresses)

✓ Management of the transition from IPv4 to IPv6

Governance of the Internet

Principles, standards, rules and procedures for decision-making

✓ Development and application of common principles, standards, rules, decision-making procedures and programs to shape the evolution and use of the Internet

✓ Technical Administration

✓ Policy Governance / Policy Direction[185]

✓ Control over the technologies used in the Internet

✓ Control over policies supporting the Internet

✓ Policy issues

✓ Standardization aspects

Stakeholders in the Internet Governance

✓ ICANN (Internet Corporation for Assigned Names and Numbers)

- o Governments
- o Civil society
- o Regional Internet Register
- o gTLD registries

184 Qui le fait fonctionner : Le écosystème Internet https://www.internetsociety.org/fr/resources/doc/2014/makes-internet-work-internet-ecosystem

185 Les acteurs de la Gouvernance de l'internet – https://www.afnic.fr/medias/documents/afnic-dossier-gouvernance-internet-06-2008.pdf

- o CcTLD registries
- o Regional Registers Organizations
- ✓ IGF (Internet Governance Forum)
  - o International Telecommunication Union (ITU)
  - o UNESCO
  - o UNDP
  - o Private sectors
  - o Scientific Community
  - o Internet Engineering Task Force (IETF)
  - o AFNIC registries
  - o Parliaments
- ✓ ISOC (Internet Society)
  - o National Chapters
  - o Charitable projects, patronage, foundations ...
  - o ISO (International Standardization Organization)
  - o Operators
  - o Companies
  - o Institutions[186]

Stakeholders of Internet Management

- ✓ Governments
- ✓ Business Sector
- ✓ Civil society
- ✓ Technical Community
- ✓ Academic sector

Internet business sector

4 fundamental stakeholders

- ✓ **Domain Name Providers / Registrars**

  2 categories: Registrars and registries
  - o Sale of domain names (eg .com, .net)
- ✓ **Internet Service Provider (ISP)**
  - o Provision of Internet service (Internet connection)
- ✓ **Telecommunications Operators**
  - o Internet traffic transport
  - o Management of the Internet infrastructure
- ✓ **Content Providers**
  - o Multimedia content provider
  - o Examples: Google, Facebook, Twitter, Disney, TV stations

Categories of Internet Operators

4 categories of Internet operators

- ✓ Internet service providers: Operator offering Internet connection to customers
- ✓ Content and Service Providers: Producer of multimedia content providing services to websites (Google, Youtube, facebook, twitter, TV channels, etc.)
- ✓ Content Delivery Networks : Set of servers, called relay servers, installed at different points of the Internet, and each containing a copy of the same content distributed to users of the network.
- ✓ Objective: facilitating rapid access during searches
- ✓ IP Transit Operator: Operator providing the link (provision of the signal transport service) between several Internet operators

Internet Services

- ✓ Platform of all kind online services (Multiplicity of digital services)
- ✓ Virtual access to a set of information and resources
- ✓ Remote execution of a set of activities once done traditionally
- ✓ Transfer of all traditional activities in the digital space

Categories of services provided by the Internet

4 main categories

Category 1: Communication Services

Different forms of communication between users

- ✓ **Email:** Global exchange of messages by electronic

186 Les acteurs de la Gouvernance de l'internet – https://www.afnic.fr/medias/documents/afnic-dossier-gouvernance-internet-06-2008.pdf

means between two or more users each provided with an email address (messages exchanged via e-mail: texts, images, videos, sound, etc.)

- ✓ **Telephony on the Internet:** Telephone conversation from anywhere between Internet users with appropriate terminals and compatible software
- ✓ **Forum:** Space allowing Internet users to exchange and share their points of view on a given subject.
- ✓ **Internet chat:** Protocol allowing Internet users to exchange texts accompanied by videos in real time between them (eg: yahoo messenger, msn messenger)
- ✓ **Instant messaging:** Service allowing the sharing of information in real time between two users (Facebook Messenger, Snapchat, WhatsApp)
- ✓ **Mailing list:** Service allowing groups of users to share common information via email
- ✓ **Access to remote servers (Telnet):** Connection to another computer connected to the Internet.

Category 2: Information Search Service

Several ways to access information on the Internet

- ✓ **File transfer**: Protocol for transferring files from one computer to another computer over the Internet
- ✓ **Archie:** Tool (considered as the first search engine on the Internet) indexing the different contents of FTP sites to allow quick access to files by their names
- ✓ **Gopher:** Tool (program) designed for distribution, search and access to documents on the Internet
- ✓ **VERONICA**: Tool (search program) to access content saved on Gopher servers

Category 3: Web Services

- ✓ Exchange of services between applications on the web (interactions: communication and exchange of data between web applications through the Internet) thanks to a set of widely used Internet protocols such as: XML, http, etc.
- ✓ Software system designed to support interoperable machine-to-machine interactions on a network
- ✓ Services offered via the web
- ✓ Software systems allowing interoperability between multiple software systems (agents) on a computer network[187]
- ✓ Examples of web services
    - o Translation of a text into another language
    - o Search on a postal code
    - o Search on a cooking recipe
    - o Conversion of one currency into another

Category 4: World Wide Web (WWW or W3)

- ✓ Access to multimedia information sites (texts, graphics, sound, videos, hyperlinks) available on several servers on the Internet thanks to browsers such as Internet explorer, Google chrome, Firefox
- ✓ World Wide Web
- ✓ Hypertext system operating on the Internet
- ✓ Network of computers serving pages linked together by hypertext links[188]

Uses of the Internet

- ✓ Remote communication
- ✓ Software sharing
- ✓ Exchange of opinions on subjects of common interest
- ✓ Display of information of general interest
- ✓ Promotion of organization
- ✓ Product promotion, and feedback gathering
- ✓ Customer Support Service
- ✓ Online Newspapers, magazines, encyclopedia and dictionaries
- ✓ Online shopping
- ✓ Global Conference[189]

187 Introduction aux Web Services https://benoitpiette.com/labo/introduction-aux-web-services.html#page2

188 Dictionnaire Sensagent http://dictionnaire.sensagent.leparisien.fr/www/fr-fr/

189 Internet

✓ Online course

Main areas of Internet use

✓ Communication

- o Chat
- o Video conference
- o Email
- o Social networks

✓ Research

- o Information
- o Books
- o References

✓ Education

- o Books
- o Reference books
- o Online Help Center
- o Expert points of view
- o Online course

✓ Financial Transactions

- o Access to online bank accounts
- o Online purchase and sale

✓ Real time updates

- o News updated
- o events in progress around the world

✓ Hobbies

- o Access to favorite songs and videos
- o Watching of movies
- o Online games
- o Chat with contacts

✓ Online booking

- o Booking of bus, plane and train tickets online
- o Simplification of the booking process

✓ Job search

- o Publication of vacancies through websites
- o Reception of job vacancy notifications via e-mail
- o Online Interview

✓ Blogging

- o Publication of personal diaries via specific websites
- o Publication of written works via specific websites

✓ Purchase

- o Online shopping
- o Home delivery as quickly as possible[190]

https://fr.slideshare.net/rgtoughracer/ppt-on-internet

Web 2.0

✓ Web evolution (World Wide Web) towards interactivity with the user

✓ Set of innovative facilities and interfaces allowing the user to interact with the web

✓ New version of the World Wide Web offering more innovation, exchanges and collaborative sites

✓ Evolution of the web and uses

Examples of Web 2.0

✓ Social networks (Facebook, Instagram)

✓ Exchange platforms

✓ Collaborative sites[191]

✓ Blogs

✓ Wikis

Web 2.0 applications

✓ Sharing of information (Rss, Tags)

190 Top 10 uses of the Internet http://top-10-list.org/2013/06/22/top-10-uses-of-internet/?utm_source=feedburner&utm_medium=feed&utm_campaign=Feed%3A+top-10-list%2FIAGi+(Top+10 + List)

191 Web 2.0 : définition, traduction - http://www.journaldunet.com/business/dictionnaire-du-marketing/1198353-web-2-0-definition-traduction/

- ✓ Sharing of images and videos
- ✓ Sending of invitation to contacts
- ✓ Online meeting
- ✓ Personal space management on a website

Browser or Navigation Software

Browser, explorer

- ✓ Software intended to allow access to the Internet
- ✓ Interface between the user and the Internet
- ✓ Functions: Website Display, Download, Search

Frequently used browsers

- ✓ Internet explorer
- ✓ Google chrome
- ✓ Mozilla Firefox
- ✓ Safari
- ✓ Opera

Internet access (terminal side)

- ✓ Computer
- ✓ Modem
- ✓ Navigation software
- ✓ Internet connection

Internet access provider

Types of Internet Service Providers (Internet Operators)

- ✓ Internet service provider
- ✓ Mobile operator
- ✓ Cable operator
- ✓ Satellite telephone operator
- ✓ Satellite Internet Service Provider

Internet access means

- ✓ Free Wi - Fi (in some places)
- ✓ Access (work, schools, universities, public places, etc.)
- ✓ Daily rate (quantity of MB or GB per day)
- ✓ Monthly subscription

Terminal equipment

- ✓ Terminal equipment: Computers, telephones, digital Tablets, TV Receiver, Modem
- ✓ Router (equipment used to connect multiple computers and other devices to a single Internet connection)

Service Availability

- ✓ Speed (bit rate)
- ✓ Congestion
- ✓ Reliability (measures against signal interruption)

Different types of connection

- ✓ Fixed access (wired and wireless)
- ✓ Mobile access

INTERNET ACCESS MODES

Access for residential subscribers, institutions, companies and cellular subscribers

Main means of access to the Internet

1.- Internet via dial -up (via a modem)

- ✓ Dial up: analog connection (data sent over the analog public telephone network)
- ✓ Internet access via the conventional telephone line (PSTN)
- ✓ Modem: Establishment of the dial-up connection (interface between the computer and the telephone line)
- ✓ Communication program: Instructions to Modem to dial a specific number provided by the Internet Service Provider (ISP)
- ✓ Dial up connection protocols: Serial Line Internet Protocol (SLIP) and Point to Point Protocol (PPP)
- ✓ Very low bit rate: 2400 bit per second to 56 kilobits per second
- ✓ Occupation of the telephone line for the duration of the Internet connection (impossibility to place and receive telephone calls)

2.- Internet via ADSL

- ✓ Use of High Frequencies of an Asymmetric Digital Subscriber Line (ADSL) Connection for Internet Access

- ✓ Installation of a Modem adapted to this service
- ✓ Availability of the telephone line for telephone calls when using the Internet

3.- Internet by Cable Modem (Internet by Cable)

- ✓ Internet access via a cable operator (Use of the same cable as the television)
- ✓ Installation of a Cable Modem (Use of spaces between television channels for data transmission (Internet))
- ✓ Faster connection than dial up and DSL (DSL: Digital Subscriber Line)

4.- Internet via Optical Fiber

- ✓ High speed internet access
- ✓ Connection of the user's modem to the endpoint via an Ethernet cable

5.- Internet via Satellite

- ✓ Satellite connection (access by radio waves)
- ✓ Systems used: DSS (Digital Satellite Systems) and DBS (Direct Broadcast Satellite)
- ✓ Dish: small satellite dish installed at the subscriber's (interface between the satellite and the subscriber's computer)
- ✓ Modem: interface between computer (digital signal) and electromagnetic waves (analog signal)
- ✓ Coaxial cable: connection between the antenna and the computer
- ✓ Service available over a long distance
- ✓ Connection: One-way connection (download only) and Upload: Dial up access via an ISP over a telephone line
- ✓ Connection: two-way connection (downloading and uploading on the satellite link without a telephone line)
- ✓ Poor signal quality in case of rain (degradation due to rain)

6.- Connection via ISDN (Integrated Services Digital Networks)

- ✓ ISDN: Telephone and access to the internet simultaneously
- ✓ Typical bitrate: 64 kbps to 128 kbps
- ✓ Performance: 2 - 3 times more than the dial -up

7.- Internet by Wi- fi

At home, at work or public places

- ✓ Wi-fi: standard for wireless Internet access (at home or in public places)
- ✓ Connection by radio waves
- ✓ Compatibility needed between standard and terminals (cellular telephones, digital tablets, computers)

8. - Wireless LAN internet

- ✓ Short-distance Internet connection
- ✓ Connection by radio waves or infrared signals
- ✓ Use of an access point
- ✓ IEEE.802.11: Standard most used by WLAN

9.- Internet via mobile telephone networks *(GPRS, EDGE, 3G, 4G)*

Cellular Networks Data Service

- ✓ GPRS, EDGE, 3G, 4G: Technologies enabling mobile Internet access
- ✓ Protocol operated: WAP (Wireless Application Protocol)
- ✓ Access to this service via cellular telephones, computers, digital tablets
- ✓ Modem needed for access via computers and tablets

10. - Internet via Wimax

Wimax: Worldwide Interoperability for Microwave Access

- ✓ Wireless technology for high-speed broadband Internet access
- ✓ Standards used by this technology: 802.16d and 802.16e

11.- Internet via PowerLine Communications

- ✓ Use of the lines of the public electricity grid for the provision of access to the Internet
- ✓ Technique used: superposition of an analog signal to AC current

Cellular subscriber data account

- ✓ Availability of data service (Internet) through cellular networks (2.5 G, 3G and 4G)

- ✓ New Account offered to Cellular Subscribers for Access to the Data Service: Data Plan
- ✓ Data account accessible to all subscribers with an Internet compatible telephone.
- ✓ Payment per MB upon the exhaustion of the daily or monthly data plan for the transmission and reception of multimedia messages
- ✓ Payment by MB: Debit of the main account (first account)
- ✓ 10 - 15 GB = monthly fee for a category of users

Consumption of bits, bytes, KB, MB and GB

- ✓ Use of the Internet: Consumption of MB, GB
- ✓ Data consumption: Each exchange of information, each Internet page consulted
- ✓ Data consumption: every email sent or received, and every little message sent via social networks
- ✓ Message Exchange: MB Consumption on the Sender and Recipient's Side
- ✓ Exchange 5MB video: 5 MB on the sender's side and 5 MB on the receiver's side during downloading
- ✓ Downloading of files sent : Quantity of MB greater than or equal to the size of the message
- ✓ Debit of Data account at each transaction (the same as a bank account with each withdrawal of money)
- ✓ Sending of a simple email: 15 KB
- ✓ Sending of an email with attachments: 1 MB to 20MB on average
- ✓ 5 Web pages viewed: 25MB
- ✓ Downloading of a song or a game of a few minutes: 3 MB to 8 MB on average
- ✓ Music in streaming mode: 500 KB on average
- ✓ Video in streaming mode: 2 MB to 5 MB on average
- ✓ Profile update: 500 KB
- ✓ Continuous consumption of MB for updates between servers and the cellular telephone

Quantity of MB per user per day

- ✓ Quantity of MB per day variable and dependent on uses
- ✓ 30 to 50 MB = Insufficient daily plan for the exchange of photos and videos during a day
- ✓ Internet search = tens of MB per day for a student
- ✓ 300 MB per day = email, social networks, internet searches, browsing
- ✓ New applications = increase of the consumption

Exhaustion of monthly plan (data volume)

- ✓ *3 possible scenarios*
- ✓ Decrease of speed (continuity of navigation at no additional cost, but with a reduced speed, generally of the order of 128 kbps, well below the speed of a 3G or 4G)
- ✓ Transfer of data billed per unit (in MB)
- ✓ Blocking of the Internet connection (disactivation of the Internet connection)

Advantages of the Internet

Social Networking: Communication with Remote People

- ✓ Main social networks: Facebook, Twitter, Yahoo, Google+, Flickr, Orkut, LinkedIn

Education and technology

Significant contributions to the enrichment of knowledge by search on the Internet by search engines

- ✓ Main search engines: Google, Bing, Search, Yahoo, etc.

Hobbies

- ✓ Online television
- ✓ Online games
- ✓ Songs
- ✓ Videos
- ✓ Social Network Applications

Online services

- ✓ Availability of a set of traditional online services
- ✓ Development of new online services

Disadvantages of Internet

- ✓ *Threats to personal information*: Risks of unauthorized use of personal information (names, address, credit card number)
- ✓ *Unsolicited* mail: Receipt of unsolicited emails leading probably to the destruction of an entire system
- ✓ *Cybercrime:* Possibility of perpetrating a set of crimes in cyber space
- ✓ *Virus* attack: Possibility of virus attack for computers and other devices connected to the Internet.
- ✓ *Wrong information*: Possibility of misleading users by incorrect information posted on certain websites

Broadband

- ✓ *Broadband* : high data transfer capacity
- ✓ *Broadband* : *High speed* connection
- ✓ *Broadband* : Ability to access the Internet with speed higher than speeds provided by dial up access (64 kbps or 56 kbps)
- ✓ *Broadband* : binary speed starting at 512 kbps (minimum speed to be provided)

Need for broadband

- ✓ Internet browsing, online games, e-commerce
- ✓ Downloading of movies, heavy files, etc.
- ✓ Access to Internet TV
- ✓ Videoconference on the Internet
- ✓ Access to a web page in seconds
- ✓ Real time courses

Broadband Foundation

- ✓ Bandwidth
- ✓ Transmission medium (links)
- ✓ Technologies

Factors Influencing Speeds (Bitrates)

Variation of speeds (Bitrates)

Depending on the transmission protocols

- ✓ ATM used by operators to convey IP
- ✓ ATM speed different from the IP speed actually used by the home-based customer
- ✓ IP speed lower than ATM speed (20% less than 25% ATM throughput)
- ✓ Example: 24 Mbps for ATM speed = 18 Mbps IP speed

Depending on the nature of the data exchanges

- ✓ Less time for downflow than upflow
- ✓ Especially applicable for an ADSL connection

Depending on the geographical location

- ✓ Decrease in speed with distance (in the case of ADSL Internet access)

Depending on the equipment

- ✓ Power of the computer
- ✓ Presence or absence of antivirus and firewalls[192]

Types of Broadband or High-Speed Connections

Different broadband transmission technologies

- ✓ Digital subscriber line (tens of Mb/s)
- ✓ Modem cable (speed exceeding 1.5 Mb/s)
- ✓ Optical fiber (hundreds of Mb/s)
- ✓ Wireless link (a few hundred Kb/s)
- ✓ Satellite (a few tens of Kb/s)
- ✓ Broadband over power lines

Broadband value chain

- ✓ Telecommunications networks
- ✓ Services provided
- ✓ User terminal
- ✓ Applications
- ✓ Contents

Means of access to High Speed Internet (in fixed mode)

Several options

- ✓ Optical fiber up to the subscriber
- ✓ ADSL from the telephone network

192 Guide pratique http://www.mediateur-telecom.fr/ressources/media/files/Guide_pratique_chapitre03.pdf

- ✓ Cable via DOCSIS technology
- ✓ 4G/LTE
- ✓ Wimax
- ✓ Satellite
- ✓ Light Fidelity (Li-Fi)

Access to high-speed Internet (in mobile mode)

Several options

- ✓ Wimax
- ✓ 4G/LTE
- ✓ Satellite
- ✓ Internet Balloons
- ✓ Drone (Internet Drone)

Digital value chain and investments

- ✓ Terminals: investment in telephones, computers, digital tablets, routers
- ✓ End User: Investment in last mile connectivity
- ✓ Telecommunications networks: investment in national and international network infrastructure
- ✓ Platform: investment in data centers and related equipment
- ✓ Digital content and applications: investment in software development and content production[193]

Downloading

- ✓ Loading of a file from a remote computer or a remote server to a local computer via a transmission channel (Intranet, Internet)
- ✓ Transfer of data from a server to a terminal (computer, tablet, phone, etc.) via Internet
- ✓ Process for having a copy of a file hosted on a remote server on a local computer
- ✓ Examples of Downloading
- ✓ Loading a local computer or phone from a file hosted on a remote server
- ✓ Loading of videos received via WhatsApp to a cellular telephone

Uploading

- ✓ Operation of loading a file from a local computer to a computer or a remote server via a transmission channel (Intranet, Internet)
- ✓ Transfer of data from a terminal to a remote server via the Internet
- ✓ Process for depositing a file in a remote server
- ✓ Uploading Examples
  - o Loading from a computer or telephone pictures on a facebook account
  - o Loading of file (attachment) when sending an e-mail

Website

- ✓ Set of web pages interlinked together by hypertext links
- ✓ Set of pages accessible by an address web
- ✓ Virtual space hosting information and services
- ✓ Virtual space allowing access to information and remote services
- ✓ Interactive communication and exchange tool
- ✓ Set of web pages viewable in a browser
- ✓ World Wide Web: set of Internet sites

Web Page

- ✓ Basic unit of a website accessible by a URL (Uniform Resource Locator)
- ✓ Web consultation unit (world wide web)
- ✓ File (HTML, Javascript, PHP, MySQL) containing several other files (images, Flash animations, video, etc.)
- ✓ Web page: accessible via a browser
- ✓ Example of web pages: home page, contacts, forum

Functions or objectives of websites

5 main functions or objectives

- ✓ Representation (showcase website)
- ✓ Sale (E-Commerce, Online Store)

193 itu regional economic and financial forum for telecommunications / ICTs for arab region - https://www.itu.int/en/itu-d/regional-presence/arabstates/documents/events/2015/eff/pres/maaref%20ott% 20presentation% 20manama% 202015.pdf

- ✓ News (news site, blog)
- ✓ Sharing (Social networks, forum)
- ✓ Work (Web Applications)[194]

Categories of websites

- ✓ "Showcase" site: digital showcase of the company or institution
    - o Presentation of the company and its services
    - o Visibility on search engines
- ✓ E-commerce website: online store
    - o Referencing of products for direct purchase on the Internet
- ✓ Event website or temporary website
    - o Communication on a particular event
    - o Type of electronic flyer
- ✓ Business Blog
    - o Space for publishing articles on products[195]

Types of website

2 types of websites

- ✓ *Static website*: Website displaying only the information (impossibility to place a request)
- ✓ *Dynamic website*: Website designed to answer visitors' requests (interactions between visitors and website databases)
    - o Example of dynamic site: www. itu.int

Main steps in the creation of website

- ✓ Reservation of the domain name (unique website access address)
- ✓ Website development (Coding in HTML, Javascript, PHP, MySQL ...)
- ✓ Hosting of the website on a web server (server space made available to the website)
- ✓ Web site launch (Launch of the website on the web for consultation)

194 Créer un site Internet – https://www.ideematic.com/dictionnaire-web/creer-un-site-internet-professionnel

195 Qu'est-ce qu'un site web ? – https://www.petite-entreprise.n - https://www.definitions-marketing.com/definition/blog/et/P-2823-85-G1-definition-qu-est-ce-qu-un-site-web.html

Blog or Web log

- ✓ Website or part of a website used for periodic and regular publication of personal articles[196]
- ✓ Personal or business page containing notices, links or columns periodically created by his or her authors in the form of posts[197]
- ✓ Online Journal or personal register (weblog)

Applications or application softwares

- ✓ Software automating the principles of an activity
- ✓ A more or less complex program, installed on a user's computer, to obtain a range of local services or through a network[198]
- ✓ Software developed to perform a task or set of tasks in a given field
- ✓ Interactive program (application software hosted on a server) accessible on the Internet via a web browser

Examples of applications

- ✓ Facebook
- ✓ Youtube
- ✓ Intagram
- ✓ WhatsApp
- ✓ Pandora
- ✓ Google Gmail
- ✓ Webmail
- ✓ Search engine

Web application

- ✓ Any application using an Internet browser as a client
- ✓ Application software hosted on a server and accessible via a web browser[199]
- ✓ Application executable through an Internet

196 Blog - https://en.wikipedia.org/wiki/Blog

197 Définitions Marketing – https://www.definitions-marketing.com/definition/blog/

198 PROGRAMME, *informatique* – https://www.universalis.fr/encyclopedie/programme-informatique/

199 Application web – https://www.ideematic.com/dictionnaire-web/application-web

browser[200]

- *Examples* :Office 365, Google sheets, Omni focus, Agile CRM, Acuity scheduling, HubSpot, Hulu, Pandora, Meebo, Google Apps, Microsoft Office Live, WebEx WebOffice, webmail, Google

Mobile application

- ✓ Downloadable program for free or paid and executable from the operating system of a smartphone or a tablet[201]
- ✓ Application software developed for mobile devices (Cellular telephones, digital tablets, PDAs, etc.)
  - Examples: Documents to Go, Dropbox, Evernote, Pandora, Snapchat, Facebook, Google Play, Google Search

Main operating systems for mobile applications

- ✓ Android
- ✓ IOS
- ✓ Windows phone

Social networks

- ✓ Websites dedicated to the online establishment of communities of people having shared or common interests
- ✓ Information sharing, person-to-person interactions, and shared and collaborative content creation

Purpose of social networks

- ✓ Facilitation of social exchanges between users

Uses of social networks

- ✓ Personal communication (informal communication channel allowing contact with family and friends any time
- ✓ Search for old friends and colleagues
- ✓ Search for new contacts
- ✓ Promotion of personal activities
- ✓ Discussion tools
- ✓ Publication tools

200 Définitions Marketing – www.definitions-marketing.com/definition/web-application/

201 Définition : Application mobile – https://www.definitions-marketing.com/definition/application-mobile/

Benefits of social networks

- ✓ Easy and practical Communication
- ✓ Connection with people around the world
- ✓ Increase of business contacts

Disadvantages of social networks

- ✓ Less face-to-face communication
- ✓ Decreased level of privacy
- ✓ Exposure to Hacker attacks
- ✓ Less time for other activities

Opportunities of social networks

- ✓ Solutions (through forum)
- ✓ Information search
- ✓ Reputation (reputation management)
- ✓ Marketing (tool to reach a general public)

Threats of social networks

- ✓ Security: risk of dishonest use of information disseminated
- ✓ Safety: problems causing harm to other users
- ✓ Privacy: Sharing of more information than recommended
- ✓ Data Integrity: Loss of Information
- ✓ Judgment from words and comments, images or videos posted
- ✓ Scam of other users

Potential consequences of the use of social networks

- ✓ Disciplinary actions of schools and universities against pupils and students
- ✓ Harassment
- ✓ Use of information posted as evidence in legal proceedings
- ✓ Identity theft
- ✓ Refusal criteria for an internship or a new job
- ✓ Disciplinary actions of an employer, (including revocation)

Over - The Top Service (OTT)

Over - The Top service: bypass service

- ✓ Application or service providing a product over the Internet and bypassing traditional distribution
- ✓ Use of existing telecommunications network infrastructures to reach the final consumer without paying anything
- ✓ Multimedia signal delivery service (transport of audio, video and data streams) through the networks of traditional operators (cable operators, telephone or satellite operators)
- ✓ Transport of video, audio or data streams over the Internet without the necessary intervention of an operator

Consequences of OTT

- ✓ No involvement of the traditional operator in the control and delivery of the service
- ✓ Bypass service: signal transport provided by the traditional operator
- ✓ Applications creating value over traditional networks without paying financial compensation
- ✓ Threat to traditional operators (replacement of the TV provider by Netflix, and replacement of the long-distance operator by Skype)
- ✓ Loss of income for traditional operators
- ✓ Continued investment in the transmission network to increase bandwidth for the provision of application services
- ✓ OTT Providers: Skype, Viber, Whatsapp, Imo, Tango)

OTT base

- ✓ User Terminals
- ✓ Data plan
- ✓ Digital Technologies

Types of services provided by OTT

3 Types of service

- ✓ Voice and messaging (VoIP, Skype, chat with and without video, Gmail, WhatsApp, Wechat, Viber, etc.)
- ✓ Appalications (Facebook, Linkedin, Twitter, Instagram, WeChat, etc.)
- ✓ Audio and video content (TV - OTT, Video - OTT, streaming and video on demand, Netflix, Netmovies, Hulu, Cuevana TV, Youtube)[202]

OTT Services

- ✓ Voice services
- ✓ Messaging services
- ✓ Teleconferencing Services
- ✓ Video streaming
- ✓ Video on demand

OTT players

Main Players

- ✓ Google
- ✓ Facebook
- ✓ Microsoft
- ✓ Yahoo!

Examples of OTT services

- ✓ Skype
- ✓ Viber
- ✓ WhatsApp
- ✓ Chat On
- ✓ Snapchat
- ✓ Instagram
- ✓ Kik
- ✓ Google Talk
- ✓ Hike
- ✓ Line
- ✓ WeChat
- ✓ Tango

Stakes of OTT

- ✓ Competition with existing Operator / ISP offers

202 Consultation Paper On Regulatory Framework for Over-the-top (OTT) services – http://www.trai.gov.in/WriteReaddata/ConsultationPaper/Document/OTT-CP-27032015.pdf

- ✓ Consumption of bandwidth of existing operators
- ✓ Creation of "value" on operator networks without their agreement
- ✓ No repayment of the financial contribution
- ✓ Revision of the business model by telecommunications operators

Strengths of OTT

- ✓ Cost
- ✓ Convenience
- ✓ Features offered
- ✓ Social trend
- ✓ Content availability
- ✓ Penetration of smart telephones and the mobile Internet
- ✓ User experience
- ✓ Neutrality of the Internet[203]

Weaknesses of OTT

- ✓ Quality of service (service dependent on the existing Internet connection)
- ✓ Availability (Internet dependency)
- ✓ Service mobility related to operator coverage

INTERNET OF THINGS (IOT)

- ✓ *Global infrastructure of the information society, enabling advanced services by interconnecting objects (physical or virtual) through existing or evolving interoperable information and communication technologies*[204]
- ✓ Extension of the Internet to things and places
- ✓ Physical objects with digital identities to facilitate communication between them[205]
- ✓ Exchange of information and data between physical objects and the Internet
- ✓ Grouping of all communicating physical objects with a unique digital identity[206]
- ✓ Direct and standardized digital identification (IP address, smtp, http protocols) of a physical object using a wireless communication system (RFID chip, Bluetooth or Wi-Fi.)[207]
- ✓ Ability of connected objects to send data reports to their owners over the Internet[208]

Principles of the Internet of Things

- ✓ Equipment of objects with communication means (Wi-Fi chip, RFID tag ... ..)
- ✓ Transmission of data and messages by objects connected to servers via the Internet
- ✓ Collection, storage, processing, visualization and analysis of received data

Communication in the Internet of Things environment

- ✓ Communication at any time
  - o On the move
  - o Night
  - o Daytime
- ✓ Communication in any place
  - o Outdoors
  - o Indoors (away from a computer)
  - o From a computer
- ✓ Communication with any object
  - o Between computers
  - o From person to person without a computer
  - o From person to object, using generic equipment

---

203 Services on Telecoms Service Providers - http://www.indjst.org/index.php/indjst/article/viewFile/62238/48529

204 Présentation générale de l'Internet des objets - Série y: infrastructure mondiale de l'information, protocole internet et réseaux de prochaine génération - Réseaux de prochaine génération – Cadre général et modèles architecturaux fonctionnels - Recommandation UIT-T Y.2060

205 Internet des objets – http://www.futura-sciences.com/tech/definitions/internet-internet-objets-15158/

206 Internet of Things - http://www.futura-sciences.com/tech/definitions/internet-internet-objets-15158/

207 Internet of Things - http://www.futura-sciences.com/tech/definitions/internet-internet-objets-15158/

208 Internet des objets: des applis nouvelle génération reliées aux objets physiques – https://fr.softonic.com/articles/internet-des-objets-applications-nouvelle-generation

- o Thing to thing [209]

Examples of connected objects

- ✓ Refrigerator
- ✓ Lamps
- ✓ Watches
- ✓ Heaters
- ✓ Smoke detectors
- ✓ Cameras

Means of connection of objects

- ✓ RFID (Radio Frequency - Identification)
- ✓ Wi -fi (Wireless Fidelity)
- ✓ Bluetooth
- ✓ SIGFOX
- ✓ Zigbee
- ✓ NFC (Near Field Communication)
- ✓ USB (Universal Serial Bus)
- ✓ Ethernet
- ✓ Z-Wave
- ✓ 2G, 3G, 4G, LTE, 5G

Main stakeholders of the Internet of Things

5 main Players

- ✓ Intel
- ✓ IBM
- ✓ Microsoft
- ✓ Google
- ✓ Cisco

CLOUD COMPUTING

Terms used: Cloud, Cloud Computing

- ✓ *Cloud computing*: Remote access via Internet to computing resources (database, computer servers, storage capacity, etc.) relocated
- ✓ Provision of services hosted on the Internet
- ✓ New way of delivering computing resources
- ✓ Provision of sophisticated computing services by virtual computing systems
- ✓ Ubiquitous, convenient, and on-demand access to a shared network and a set of configurable computing resources (such as: networks, servers, storage, applications, and services)[210]
- ✓ Use of computing power or storage of remote computer servers via a network, usually the Internet[211]
- ✓ Access from anywhere via a terminal (computer, cellular telephone, digital tablet) or files stored in a remote server
- ✓ Provision of sophisticated set of services accessible from anywhere with a set of hardware, network connections, and softwares[212]
- ✓ Use of memory and computing capabilities of computers and servers spread all over the world and linked by a network, such as the Internet[213]

Consumer Use of Cloud Computing Services

- ✓ Storage and sharing of data (documents, photos, videos, music) on Dropbox, Google drive, Microsoft OneDrive, Apple icloud
  - o Opening an account
  - o Allocation of storage capacity (free or payment)
  - o Access from anywhere to documents via any terminal
  - o Addition and deletion of files from anywhere

Cloud Computing Service Categories

3 categories of service

209 Présentation générale de l'Internet des objets – Série y: infrastructure mondiale de l'information, protocole internet et réseaux de prochaine génération - Réseaux de prochaine génération – Cadre général et modèles architecturaux fonctionnels - Recommandation UIT-T Y.2060

210 C'est quoi le cloud ? – http://www.culture-informatique.net/cest-quoi-le-cloud/

211 Cloud computing – https://fr.wikipedia.org/wiki/Cloud_computing
Cloud Computing: Principles and Paradigms, John Wiley & Sons, 2010

212 Cloud computing: Principles and Paradigms, John Wiley & Sons, 2010

213 Le 5e écran: les médias urbains dans la ville 2.0 – https://books.google.ht/books?isbn=2916571264

Infrastructure as a Service

- ✓ Access to a physical or virtual computing platform for performing tasks
- ✓ Remote access to a virtual computer pool (access to computing resources in a virtualized environment)
- ✓ Services available: server space, network connections, bandwidth, IP addresses, load balancers, etc.
- ✓ Ability for users to install on virtual machines
  - o Operating system
  - o Applications
- ✓ *Examples of use*: Lease of processing, storage, network and other computing resources on a remote server for operations management
- ✓ *Examples of providers*: Amazon EC2, Windows Azure, Rackspace, Google Compute Engine, Datapipe, Gogrid, Navisite, Savvis, Verizon, Rackspace, Opsource, Cloudscalling, etc.

Platform as a Service (Paas)

- ✓ Access to a computing platform (operating system and infrastructure) for the execution of specific tasks (operations management)
- ✓ Remote access to a complete computer platform (operating system, programming language, runtime environment, database, web server, etc.)
- ✓ Creation and development of Web applications
- ✓ Access to operating systems, applications and remote infrastructure
  - o Control of applications by user
  - o Addition of other tools by user
- ✓ *Examples of use*: Deployment of applications on a remote platform
- ✓ *Examples of providers*: AWS Elastic Beanstalk, Windows Azure, Heroku, Force.com, Google App Engine, Apache Stratos, Engine Yard, Appfrog, Amazon Aws, Active State, 10gen, etc.

Software as a service (Saas)

- ✓ Use of applications over the Internet
- ✓ Teamwork
- ✓ Applications (application software) at the service of users
  - o Manipulation of applications via a web browser
  - o (no installation, set up and running for the user)
- ✓ *Examples of use*: Gmail, outlook, office 365, Google apps, webmail, etc.
- ✓ *Examples of providers*: Google Apps, Microsoft office 365, box, salesforece.com, Concur, Docusign, Dropbox, Slack, Zendesk, etc.

Elements of the Cloud Computing Environment

- ✓ Infrastructure
- ✓ Platforms
- ✓ Softwares[214]

Benefits of cloud computing

- ✓ Speed of data processing
- ✓ No acquisition of hardware and software
- ✓ Access to services from anywhere
- ✓ Data security
- ✓ Less expense (no computing infrastructure, no software purchase)

Disadvantages of cloud computing

- ✓ Possibility of non-confidentiality
- ✓ Exposure to cyberattack

Digitization of traditional services

- ✓ Use of digital technologies to make traditional services (public and private) available online
- ✓ Access to public information online
- ✓ Renewal of passport online
- ✓ Online Voting
- ✓ Online banking
- ✓ Online purchase
- ✓ Online ticket booking
- ✓ Online bill payment

214 Le 5e écran: les médias urbains dans la ville 2.0 – https://books.google.ht/books?isbn=2916571264

- ✓ Online games
- ✓ Etc.

Electronic version of almost "everything"

Online, Tele, electronic, digital, (e-) and cyber: Five terms to go to the digital universe

An electronic or digital version for all traditional practices

- ✓ Digitization of traditional activities for use in the cyberspace
- ✓ Classroom course/Online course
- ✓ Traditional games/Online Games
- ✓ Traditional Services/Online Services
- ✓ Traditional Radio/Online Radio, web radio
- ✓ Traditional mail/Electronic mail (e-mail)
- ✓ Governance/e - Governance
- ✓ Visa/evisa
- ✓ Traditional Vote/Electronic vote
- ✓ Commerce/Electronic Commerce
- ✓ Traditional government/Electronic Government
- ✓ Education/ tele education, online education
- ✓ Medicine/Telemedicine
- ✓ Marketing / Telemarketing
- ✓ Coffee / Cybercafe
- ✓ Health / Cyber health
- ✓ Security / Cyber security
- ✓ Crime / Cybercrime
- ✓ Physical space / Cyberspace

CHAPTER 8

ECONOMICS OF TELECOMMUNICATIONS AND ICTs

Transversality of Telecommunications Services: Engine of the Global Economy

ELEMENTS OF A TELECOMMUNICATIONS / ICTs MARKET

- ✓ Number of fixed and mobile operators
- ✓ Number of Internet Service Providers (ISPs)
- ✓ Number of mobile virtual network operators (MVNO)
- ✓ Number of radio stations
- ✓ Number of free – to – air and scrambled television stations
- ✓ Number of cable operators
- ✓ Number of fiber optic infrastructure operators
- ✓ Active Internet Domain names
- ✓ Number of leased lines
- ✓ Contribution of the sector to the economy
- ✓ Revenues
- ✓ Investments in the sector
- ✓ Number of telephone lines and Penetration Rate (Fixed and Mobile Line Penetration, Internet)
- ✓ Number of Internet Exchange Points (IXPs)
- ✓ Length of Fiber Optic Deployed (In km)
- ✓ International Bandwidth (Gbps)
- ✓ Organization of the sector (Ministry, Regulator, Frequency national agency)

Economics of telecommunications and ICTs sector

- ✓ Investment in infrastructures of telecommunications
- ✓ Acquisition of telecommunications terminals
- ✓ Cost of operator licenses
- ✓ Acquisition of technologies
- ✓ Cost of operation of telecommunication systems
- ✓ Consumption of telecommunication services
- ✓ Etc.

Economic aspects of telecommunications / ICTs

- ✓ Production
  - o Products
  - o Services
  - o Equipment
  - o Infrastructures
  - o Systems
- ✓ Consumption
  - o Products
  - o Services
- ✓ Economic regulation
  - o Cost of services and products
  - o Cost of operating licenses
  - o Cost of operations
- ✓ Costs related to the use of the resources of the sector
  - o Radio frequency spectrum
  - o Numbering plan
  - o Internet domains
  - o High points (Designated points for transmission)
  - o Existing infrastructures (transmission network, power grid, telecommunications towers)
  - o Orbital positions
- ✓ Use of other resources
  - o Electrical energy
  - o Space rental
- ✓ Democratization of informatics and broadband
- ✓ Rates insensitive to time and distance

Economic players in the Telecom / ICTs sector

- ✓ Consumers
- ✓ Manufacturers
- ✓ Service providers

- ✓ Standardization agencies
- ✓ Regulators
- ✓ Content providers
- ✓ Software and application developers

Economic impacts of the sector

- ✓ Generation of high-income jobs
- ✓ Significant contribution to GDP
- ✓ Stimulation of productivity and GDP growth
- ✓ Support to the creation of high-growth businesses
- ✓ Key source of competitive advantages
- ✓ Creation of fields of activities and new methods of operating businesses
- ✓ Stimulation of innovation[215]

Liberalization of the telecommunications sector

- ✓ Free access to the telecommunications market
- ✓ Opening of the market to other telecommunications operators
- ✓ End of the monopoly of an administration on an activity defined by the public authority
- ✓ Possibility of intervention in the market offered to other actors

Advantages of liberalization of the telecommunications sector

- ✓ Stimulation of competition
- ✓ Technological developments
- ✓ Access to services for all
- ✓ Better quality of service
- ✓ Lower costs for users

Process of liberalization of the telecommunications sector

- ✓ Assessment of the law governing the sector
- ✓ Promulgation of new laws adapted to the development of the sector
- ✓ Privatization of the incumbent operator (operator holding the monopoly on telecommunications services)
- ✓ Liberalization of the installation of the subscriber (Customer Premises Equipment: telephones, fax, etc.)
- ✓ Grant of concession to new fixed and mobile operators
- ✓ Liberalization of the international gateway

Entry of an operator in the telecommunications market

- ✓ Call for tender
- ✓ Expression of interest (submission of a project)
- ✓ Acquisition of an existing operator by a new operator or a foreign operator
- ✓ Merging of a local operator and a foreign operator

Steps for the provision of telecommunication services in a market

- ✓ Feasibility study
- ✓ Market study
- ✓ Procurement of License or concession for operation
- ✓ Deployment of infrastructure
- ✓ Testing
- ✓ Launch of services

Licensing and concessions for a telecommunications operator

Key factors in grant of licenses and concessions to telecom operators

- ✓ Market size
  - o Population
  - o Targeted coverage zones
  - o Potential customers
- ✓ Economic Situation
  - o Global economic situation
  - o Purchasing power of consumers
- ✓ Types of services to be provided

215 The economic benefits of Information and Communications Technology - Just the FACTS http://www2.itif.org/2013-tech-economy-memo.pdf

- o Mobile telephony / 2G, Mobile telephony / 3G, Mobile telephony / 4G and 5G
- o Data transmission
- o Internet
- o Broadcasting operator (sound)
- o Television operator (Free to air or scrambled tv stations)
- o Etc.

Economics of telecommunication networks

- ✓ Fixed service price
- ✓ Mobile service price (higher due to the amount of equipment deployed to make the service available over a larger area and the user's route)
- ✓ Direct payment of services (with the credit on the user's account)
- ✓ Postpaid services (services paid after consumption, monthly subscription)
- ✓ OTT services paid in MB (Bit, Byte, Kilobyte, Megabyte)

Components of the Telecommunications Market

- ✓ Fixed telephony
- ✓ Mobile telephony
- ✓ Data transmission
- ✓ Internet
- ✓ Television

Costs borne by the consumer

- ✓ Allocation of a monthly budget for the consumption of telecommunications services of all kinds
- ✓ Continued expenditures in the use of services on a daily basis
- ✓ Expenditure for the acquisition of the service and the product (Terminal)
  - o Service: access and use of an intangible "service" (telephone call, SMS exchange, sending of electronic mail, etc.)
  - o Product: telephones, radio and television receivers, modems, digital tablets
  - o Installation of equipment

Costs borne by telecommunications operators

- ✓ Investment in network infrastructures
- ✓ Operational cost of telecommunications networks
- ✓ Costs of operating licenses

Accounting of telephony customers

- ✓ For fixed telephony: Number of lines providing a bit rate of 64kps (or 56kbps for North American operators)
- ✓ For mobile telephony: Number of SIM cards in circulation = number of mobile customers
- ✓ For telephony VoIP: Number of subscribers with a bit rate greater than 128Kbps

TELECOMMUNICATIONS MARKET (FIXED AND MOBILE TELEPHONY, INTERNET)

- ✓ Revenues (Turnover): millions, billion dollars and euros
- ✓ Millions of minutes and SMS
- ✓ Fixed and mobile subscribers
- ✓ Outgoing and incoming traffic
- ✓ Average revenue per minute
- ✓ Monthly average Income
- ✓ Average revenue per user
- ✓ Incoming call termination rate
- ✓ Roaming service

Fixed telecommunications market

- ✓ Optical fiber connection
- ✓ DSL connection
- ✓ Transition towards the new generation of networks

Mobile telecommunications market

- ✓ Spectrum allocation for wireless telecommunications
- ✓ Deployment of base stations

- ✓ Market shares and number of customers for each operator
- ✓ Mobile prepaid market
- ✓ Mobile postpaid market

Internet Market

- ✓ Number of Subscribers
- ✓ Penetration rate
- ✓ Average monthly invoice

Internet Usage Market

2 types

Classic uses

- ✓ Use of instant messaging
- ✓ Number of emails
- ✓ Volume of mobile data consumed

New usages

- ✓ Number of applications downloading
- ✓ Number of users of catch-up TV
- ✓ Music streaming revenue (music, video)

Market Segments of the Telecom / ICTs Sector

Three Market Segments

- ✓ Fixed Communications
  - o Voice Switching Equipment
  - o Routers and Other Equipment for Data Networks and Internet
  - o Equipment for Transmission Networks (Optical and Radio)
  - o Communication Equipment access (in particular broadband ADSL equipment)
- ✓ Mobile communications
  - o Terminals
  - o Radio access equipment
  - o Core Network equipment
- ✓ Private communications
  - o Business market
  - o Specific sectors such as space, transport

Market share calculation

- ✓ Voice services: Telephone calls
- ✓ Mobile data services: SMS, MMS (text messaging services)
- ✓ Data transmission services: frame relay networks, X.25, ATM, MAN, IP VPN and leased lines
- ✓ Internet service: Low-speed (dial-up) connection, broadband (ADSL, cable modem, FTTx ....) connection

Assessment of telephony customers

- ✓ Population covered (geographical area served)
- ✓ Number of monthly subscriptions (fixed and mobile telephony)
- ✓ Number of SIM cards in use (mobile telephony)
- ✓ Traffic carried

Telecommunications and ICTs revenues

4 Markets

- ✓ Sound and television broadcasting stations
- ✓ Telephony (fixed, cellular and IP)
- ✓ Data transmission (Internet, etc.)
- ✓ Social networks

Sources of Revenues of the Telecommunications Sector

- ✓ Installation or Connection Fees
- ✓ Equipment and Line Rentals
- ✓ Service Subscription Fees
- ✓ Call Costs (Local, National and International Calls)
- ✓ Other Operator Transit Costs[216]

Government revenues from the Telecom / ICTs sector

- ✓ License fees (license costs)
- ✓ Taxes on Turnover
- ✓ Costs on telecommunication links (frequency usage)
- ✓ Taxes on imported telecommunication equipment

216 Telecommunications Law – Ian Lloyd et David Mellor

- ✓ Telecommunication Equipment Licensing Fees
- ✓ Homologation fees for telecommunications equipment
- ✓ Costs on other resources used: telephone numbering plan, high points, existing infrastructure, internet domains, ect.

Revenues of radio and Television Stations

- ✓ Advertisings
- ✓ Sale of Airtimes
- ✓ Monthly Subscription (Cable oparator)

Funding of radio stations

- ✓ Indirect funding of radio and TV stations by listeners and viewers
  - o Advertising for products and services consumed by the users
  - o Payment by the Sponsors to radio and television stations (from sales revenue of products and services)

Fixed, cellular and IP Telephony revenues

- ✓ Installation costs
- ✓ Monthly subscriptions
- ✓ Telephone calls (Communications)

Income per user

- ✓ Income: Base of any projection on the telecom market
- ✓ Average revenue per user (ARPU)
- ✓ ARPU: Variable Indicator based on the market and other criteria
  - o Purchasing power of the Population
  - o Availability of services
  - o Level of usage of services

Revenue from Data Transmission (Internet)

- ✓ Monthly Subscription
- ✓ Revenue generated by daily Consumption of Megabytes and GigaBytes

Social media revenue

Different sources of revenue from social networks

- ✓ Advertising: promotion of services and products of companies on their platforms
- ✓ Marketing of recommendation
- ✓ Partnerships
- ✓ Branding elements
- ✓ Virtual currency
- ✓ Google Adsense (monetization program offered by Google to website publishers to generate advertising revenue from the performance)
- ✓ Premium accounts
- ✓ Membership fees: strategy practiced by some social networks
- ✓ Sale of virtual goods (digital gifts, additional services, etc.)
- ✓ Application fees: percentage in revenues generated by web applications running on their platforms
- ✓ Revenues from users (user payment for certain services such as additional image storage, etc.)
- ✓ Online games
- ✓ Etc.

Consumption indicators

- ✓ Number of traditional fixed lines
- ✓ Total number of mobile customers, number of mobile customers by 3G and LTE technology
- ✓ Number of prepaid, postpaid mobile customers
- ✓ Number of fixed Internet subscribers (total and per access technology)
- ✓ number of broadband subscribers
- ✓ Number of VoIP subscribers
- ✓ Market share of major mobile operators
- ✓ Market share of leading high-speed Internet service providers[217]

Revenue indicators

- ✓ Fixed line revenues
- ✓ Mobile services revenue, Mobile voice revenue,

217 Marché mondial des services et acteurs télécoms Tendances & Analyses, S1 2015 – www.idate.org/2009/pages/download.php?id...t...Telecoms...pdf...

Mobile data revenues

- ✓ Fixed Internet revenues
- ✓ Revenue from data services[218]

Interactions in the telecommunications market

- ✓ Between the manufacturer and the user: purchase of telecommunications terminals
- ✓ Between the user and the telecommunications operator: provision of telecommunications service to the user and billing by the operator
- ✓ Between the operator and the manufacturer: submission of the specifications from the operator to the manufacturer for equipment manufacturing
- ✓ Between the operator and the manufacturer: Supply by the manufacturer of telecommunications equipment to the operator

Competition in the telecommunications sector

Some forms of competition in the Telecommunications sector

- ✓ *Costs of services*: Reduction of rates between competitors to keep customers and attract new ones from other telephone operators.
- ✓ *New Services and Products*: Provision of new services and products to keep subscribers and attract competitors' ones.
- ✓ *Human resources*: Recruitment of more experienced employees and attraction of competitors' employees by offering better wages and working conditions
- ✓ *Covered areas*: greater coverage of the telecommunications service = greater market share for the telephone operator in question. National coverage: challenge of new entrants in the telecommunications market.
- ✓ *Technologies used*: Robustness and quality of technologies = quality of services provided. Best Technology = Best Services
- ✓ *Customer Service*: The platform's ability to answer questions and resolve subscriber issues in record time.
- ✓ *Licenses*: Determination of the position of an operator on the market by the number of operating licenses held (telephony, Internet, transport networks, digital television, etc.)

ICTs impacts on the traditional economy

- ✓ Redefinition of production, transportation and distribution processes
- ✓ Growth lever for other sectors of economic activities
- ✓ Global coverage of local economic activities through websites

DIGITAL ECONOMY

- ✓ Markets based on digital technologies
- ✓ Set of economic activities based on platforms such as: the mobile phone network, the Internet
- ✓ Fields of economic activities based on digital services (hardware, software, telecommunications, networks etc.)
- ✓ Online services: e-commerce, online games, online music, digital books, etc.
- ✓ Provision of traditional services online (through telecommunications networks)
- ✓ Invasion of the traditional economy by digital technologies
- ✓ Creation of new products and services, jobs, businesses (Google, Apple, facebook, Amazon, Microsoft)
- ✓ Creation of virtual goods and services in the digital industry
- ✓ Provision of a package of services across digital networks

Impacts of Digitization on the Traditional Economy

- ✓ Growth
- ✓ Productivity

Digital Economy Infrastructure

- ✓ Digital Infrastructure: Basis of the Digital Economy
- ✓ Digital Infrastructure: Economic Growth lever
- ✓ Digital Telecommunication Networks

218 Marché mondial des services et acteurs télécoms Tendances & Analyses, S1 2015 – www.idate.org/2009/pages/download.php?id...t...Telecoms...pdf...

o Production and distribution Support

Sources of Digital Economy Revenues

- ✓ Supply and Demand: More Services, More Products
- ✓ ICTs Investment and Revenues
- ✓ Transversality of the Digital Technologies Sector: Use of ICTs Services and Products in all other Fields of Activities

Players in the digital economy

4 main players

- ✓ Companies providing ICTs services (Telecommunications, IT, Electronics)
- ✓ Enterprises based on the emergence of ICTs (online services, video games, e-commerce, media and online content)
- ✓ Companies using ICTs in their activities to increase productivity (banks, insurance, automotive, aeronautics, distribution, administration, tourism, etc.)
- ✓ Individuals and households using ICTs services in their daily activities for leisure, culture, health, education, banking, social networks, etc.[219]

Transactions in the Telecom / ICTs sector

- ✓ Transactions based mainly on the number of users (telephone subscribers, television subscribers, connections, customers, etc.)
- ✓ Acquisition: Repurchase of an operator by another operator (absorption)
- ✓ Merger: Bringing together the operations of two telecommunications operators

Telecom / ICTs acquisition

- ✓ Cingular / ATT
- ✓ ALLTEL / WESTERN wireless
- ✓ Facebook / WhatsApp
- ✓ Microsoft / skype
- ✓ Google / Motorola mobility
- ✓ Google / YouTube
- ✓ Nokia / Alcatel –Lucent

Telecom / ICTs merger

- ✓ Sprint - Nextel
- ✓ Lucent - Alcatel

Digital Financial Services

- ✓ Provision of financial services to unbanked consumers through ICTs and non bank retailing chains
- ✓ Telecom / ICTs Uses in Finance
  - o Online Purchasing (Payment by Credit Card)
  - o Online Banking
  - o Transfer and mobile Payment

Digital Financial Services Access Means

- ✓ Terminals (telephones, tablets, computers)
- ✓ Connection to the network (telecommunications, Internet)
- ✓ Account opening

Players involved in digital financial transactions

- ✓ Telecommunications Operators
- ✓ Central banks
- ✓ Payment Platforms
- ✓ Service Providers
- ✓ Regulators
- ✓ Mobile money operators
- ✓ International Organizations (World bank, GSMA)
- ✓ Standardization agencies

Financial Transaction Issues

- ✓ Telecommunications Network Reliability
- ✓ Network Security (Cybersecurity)
- ✓ Digital identity
- ✓ Interoperability
- ✓ Data confidentiality
- ✓ Terminals and platforms user-friendliness
- ✓ Accessibility for all
- ✓ Users Trust

219 Observatoire du numérique – http://www.entreprises.gouv.fr/observatoire-du-numerique/economie-numerique

Commercial Value of the Internet

- ✓ Reduction of operating costs
- ✓ Creation of new sources of revenue
- ✓ Development of new markets and channels
- ✓ Development of new products on the Web
- ✓ Attraction of new customers
- ✓ Loyalty increase and retention of customers

CHAPTER 9

TELECOMMUNICATIONS AND NATURAL DISASTERS

EMERGENCY TELECOMMUNICATIONS

- ✓ Set of electronic communications means used in case of emergency
- ✓ Means to predict and detect disasters, and to launch alert

Importance of telecommunication services in emergency situations

- ✓ Before: Preparedness (warning to the population and prevention)
- ✓ During: Management (response and intervention of the authorities concerned, communication with relatives and parents, implementation of the means and operations of relief)
- ✓ After: Recovery (Assistance to the resumption of Activities)

Types of communication facilitated in case of emergency

- ✓ Citizen communication with authorities
- ✓ Communication between authorities
- ✓ Alerts from authorities to citizens
- ✓ Communication between affected citizens in case of disasters
- ✓ Remote monitoring system of risk situations[220]

Emergency telecommunicatios players

- ✓ Telephony, sound and television broadcasting operators
- ✓ Internet service providers
- ✓ Amateur radio
- ✓ Regulators
- ✓ International organizations (ITU, ETC, TSF)
- ✓ Elecrical power providers

Interventions of the International Telecommunication Union (ITU)

- ✓ Deployment of wireless communications systems
- ✓ Deployment of satellite communication Equipment
- ✓ Deployment of Wimax networks
- ✓ Assistance in establishing the damage assessment in telecommunications networks
- ✓ Re-establishment of telecommunication networks
- ✓ Provision of expert services for systems operation

Framework for ITU cooperation in emergencies

- ✓ Deployment of emergency telecommunication facilities for Member States
- ✓ Three-fold framework for this assistance

1. Technology Component

- o Grouping of satellite operators, land-based earth stations, telecommunications operators, mobile operators and geographic information system providers for the provision of retrospective information before, during and after disasters

2. Financial Component

- o Establishment of a reserve fund supplied by member states, development banks and regional economic groups

3. Logistics Component

- o Support service for transporting telecommunications equipment to and from the disaster sites (use of air and courier services)[221]

ITU principles for emergency communication

4 strategic principles

1. – *Consideration of all risks*
   - ✓ Natural disasters: cyclones, floods, droughts, tsunamis, fires, earthquakes
   - ✓ Man-made disasters: fires, shipwrecks

2. - *Use of all types of technologies*
   - ✓ Broadcasting
   - ✓ Radio Amateur
   - ✓ Cellular telephony
   - ✓ Internet

3.- *Interventions at all levels*
   - ✓ Prevention

220 Plan national des télécommunications d'urgence https://www.itu.int/en/ITU-D/EmergencyTelecommunications/Documents/Cameroon_2007/Presentations/Pr_MINPOSTEL.pdf

221 Télécommunications d'urgence - ITU http://www.itu.int/net/itunews/issues/2011/02/28-fr.aspx

- ✓ Preliminary planning
- ✓ Interventions
- ✓ Organization and management of relief operations
- ✓ Reconstruction of damaged telecom networks for sustainable development

4.- Involvement of all stakeholders

- ✓ Partnerships with development partners for access to ICTs for people living in rural and remote areas
- ✓ Involvement of rural communities, central governments, private sector, civil society, international organizations to contribute to the development of ICTs[222]

TAMPERE CONVENTION

- ✓ Convention concluded at Tampere (Finland, 8 June 1998)
  - o Treaty to save lives through telecommunications in case of natural disasters
- ✓ Convention entered into force in 2005 and ratified by many countries
- ✓ Increased speed and efficiency of relief
- ✓ Provision of telecommunications resources for mitigation disasters
- ✓ Provision of telecommunication resources for disaster relief operations

Obstacles before the arrival of the Tampere Convention

- ✓ Cross-border use of telecommunications equipment subject to the regulations in force in the affected areas
- ✓ Prior consent of local authorities to the import and rapid deployment of telecommunications equipment

Solutions brought by the Tampere Convention

- ✓ Mobilization of States for the rapid provision of telecommunication assistance in the event of a disaster
- ✓ Installation and implementation of reliable and flexible telecommunications services
- ✓ Simplification of the use of telecommunications equipment
- ✓ Removal of regulation obstacles for the use of telecommunication resources for disaster mitigation (no licensing requirements for the use of allocated frequencies, no restriction on the import of equipment, no limit for movements of humanitarian workers)[223]

Telecoms without borders

- ✓ Organization involved in the provision of telecommunication services in emergency situations

Main activities of Telecoms Without Borders

- ✓ Rapid Response Telecommunication Center
- ✓ Humanitarian Call Operations - Free Calls for Residents
- ✓ Local Capacity Building for Disaster Preparedness
- ✓ Reduction of the Digital Divide through Long-Term Community Centers[224]

Emergency Telecommunications Cluster (ETC)

- ✓ Global network of organizations working together to provide shared communication services during humanitarian disasters

Main activities of ETC

- ✓ Local and global coordination
- ✓ Voice and data connectivity
- ✓ Services for the communities
- ✓ Capacity building for disaster preparedness

CONSEQUENCES OF NATURAL DISASTERS ON TELECOMMUNICATION NETWORKS

3 types of consequences on telecommunications networks

1.- Physical destruction of telecommunication network infrastructures

222 Télécommunications d'urgence http://www.itu.int/fr/ITU-D/Emergency-Telecommunications/Pages/default.aspx

223 Les télécommunications sauvent des vies http://www.itu.int/itudoc/gs/promo/bdt/flyer/87636-fr.pdf

224 slideplayer.com/slide/1668520

- ✓ Network elements likely to be affected: radios (transmitters / receivers), antennas, cables, towers, etc

2.- Destruction of support infrastructure

- ✓ Dependence of telecommunications networks on electrical energy
- ✓ Destruction of the power grid: unavailability of electrical energy for the operation of the telecom networks

3.- Congestion of telecommunication networks

- ✓ High Demand for connection during and after natural disasters (almost 100% of customers)
- ✓ Failure of the network to handle all these requests and route all the resulting traffic
- ✓ Networks designed and sized for the routing of a percentage of subscribers' traffics in normal time

www.ingramcontent.com/pod-product-compliance
Lightning Source LLC
Chambersburg PA
CBHW080545220326
41599CB00032B/6367

* 9 7 8 9 9 9 7 0 5 3 0 0 8 *